Disaster Mitigation, Risk Reduction, and Resilience Design of Engineering Structures

Disaster Mitigation, Risk Reduction, and Resilience Design of Engineering Structures

Editors

Liqiang Jiang
Wei Chen
Chang He
Yi Hu
Qi Cai

MDPI

Basel • Beijing • Wuhan • Barcelona • Belgrade • Novi Sad • Cluj • Manchester

Editors
Liqiang Jiang
Central South University
Changsha
China

Wei Chen
China University of Mining and Technology
Xuzhou
China

Chang He
Central South University
Changsha
China

Yi Hu
Central South University of Forestry & Technology
Changsha
China

Qi Cai
Hohai University
Nanjing
China

Editorial Office
MDPI
St. Alban-Anlage 66
4052 Basel, Switzerland

This is a reprint of articles from the Special Issue published online in the open access journal *Buildings* (ISSN 2075-5309) (available at: https://www.mdpi.com/journal/buildings/special_issues/E66E2NQ58S).

For citation purposes, cite each article independently as indicated on the article page online and as indicated below:

Lastname, A.A.; Lastname, B.B. Article Title. *Journal Name* **Year**, *Volume Number*, Page Range.

ISBN 978-3-7258-0947-9 (Hbk)
ISBN 978-3-7258-0948-6 (PDF)
doi.org/10.3390/books978-3-7258-0948-6

Cover image courtesy of Siqi Shen

Contents

buildings

MDPI

Article

Research on Optimization Design of Prefabricated ECC/RC Composite Coupled Shear Walls Based on Seismic Energy Dissipation

Jian Yang, Ming Sun *, Guohuang Yao, Haizhu Guo and Rumian Zhong

School of Transportation and Environment, Shenzhen Institute of Information Technology, Shenzhen 518172, China; 2021000169@sziit.edu.cn (J.Y.); 2015000010@sziit.edu.cn (G.Y.); guohaizhu@126.com (H.G.); 2023000106@sziit.edu.cn (R.Z.)
* Correspondence: msunac@connect.ust.hk

Abstract: This study explores an advanced prefabricated composite structure, namely ECC/RC composite shear walls with enhanced seismic performance. This performance enhancement is attributed to the strategic use of engineered cementitious composites (ECC) known for their superior ductility. The study conducts both experimental and numerical simulation analyses to scrutinize the seismic energy absorption capabilities of this innovative structure. Emphasis is placed on critical aspects, such as the optimal deployment areas for ECC within composite coupling beams and shear walls, the grade of ECC strength, the proportion of stirrups in coupling beams, and the caliber of longitudinal reinforcement. Through finite element analysis, this research quantitatively assesses the impact of these variables on seismic energy dissipation, incorporating evaluations of load–displacement hysteretic behaviors and the energy dissipation potential of ECC/RC shear wall samples. The findings suggest the optimal ECC application in the coupling beams, and within a 14% structural height at the base of shear walls. Recommended design parameters include an ECC strength grade of E40 (40 MPa), longitudinal reinforcement of HRB400 (400 MPa), and a stirrup ratio in coupling beams of 0.5%.

Keywords: prefabricated ECC/RC shear wall; seismic energy dissipation; ECC; finite element analysis; structural optimization; coupling beams

Citation: Yang, J.; Sun, M.; Yao, G.; Guo, H.; Zhong, R. Research on Optimization Design of Prefabricated ECC/RC Composite Coupled Shear Walls Based on Seismic Energy Dissipation. *Buildings* **2024**, *14*, 951. https://doi.org/10.3390/buildings14040951

Academic Editor: Andreas Lampropoulos

Received: 25 February 2024
Revised: 15 March 2024
Accepted: 27 March 2024
Published: 30 March 2024

1. Introduction

The advancement of intelligent science and technology has become a pivotal strategy globally. As economies and societies evolve, traditional industries, characterized by high resource consumption, environmental pollution, and labor intensity, are under increasing pressure to transform and upgrade. In the construction sector, conventional on-site pouring methods are beset with challenges, including substantial resource consumption, low efficiency, resource wastage, quality assurance difficulties, and environmental degradation. To address these challenges, the industrialization of prefabricated building construction has gained increasing attention and adoption. This method represents a modern approach in the construction industry, characterized by design standardization, factory production, assembly construction, and intelligent, informatized management [1].

Prefabricated structural systems, encompassing concrete, steel, and timber structures, are extensively utilized. Among these, prefabricated concrete structures are predominantly employed in residential, educational, and office buildings. The prefabricated shear wall structure, in particular, is extensively used in high-rise buildings due to its lower resource consumption and environmental impact.

Recent research has focused on the mechanical properties, ductility, and durability of prefabricated concrete shear wall structures [2–5]. These studies have demonstrated that such structures possess an enhanced lateral stiffness and energy dissipation capacity, making them suitable for seismic regions. However, the connections between prefabricated components, especially horizontal connections, are crucial for seismic performance and warrant further attention.

Various horizontal connection methods in precast shear walls, such as grouting sleeve, constrained slurry anchor steel bar, and metal bellows slurry anchor steel bar connections, have been explored. In China, grouting sleeve connections are recommended as the most mature and effective reinforcement technology. However, these connections in prefabricated shear wall structures have shown limitations in seismic resistance compared to cast in situ structures. Peng et al. [6] conducted low-cycle repeated load tests on grouting sleeve-connected prefabricated shear wall specimens, revealing inferior stiffness and energy dissipation compared to cast in situ specimens. Brunesi et al. [7,8] investigated the seismic performance of a full-scale, two-story, lightly reinforced precast concrete wall panel structure connected by steel connectors and mortar, identifying the joints as the weakest structural elements. Similarly, studies by Qian et al. [9] and Wu et al. [10] indicated that despite mechanical similarities to cast-in-place structures, the lower ductility of joints in prefabricated structures compromises their structural integrity and robustness. Qian et al. [11] observed significant damage to coupling beams in a three-story assembled shear wall concrete structure connected by mortar sleeves, although the story drift conformed to China's high-rise building structure design code [12]. Typical earthquake damage in grouted sleeve-connected prefabricated shear wall structures includes loss of coupling beam function, concrete destruction at the compression side wall base, and reinforcement yielding or slipping out of the concrete [13–16].

Currently, prefabricated components in concrete shear wall structures are primarily used for vertical load-bearing and enclosure systems, while lateral force-resisting shear wall members are often constructed in situ. This approach results in a low prefabrication and assembly rate, complicating construction processes. To enhance the widespread adoption and application of prefabricated concrete shear wall structures, it is imperative to strengthen and guarantee their seismic performance effectively.

Designing structures for seismic resistance involves reducing seismic demand based on the ductility level permissible within the structural system. Special attention must be paid to ensuring that critical structural parts can withstand the required nonlinear deformation without significant strength loss [17]. Therefore, employing materials with high ductility or damping, such as engineered cementitious composites (ECC), in key stress and energy dissipation areas can enhance the structure's seismic energy dissipation capacity.

ECC, a material with high ductility and comparable compressive strength to concrete, exhibits exceptional tensile strain hardening capability [18]. It also has superior damping and energy dissipation capacities [19], effectively absorbing energy input into the structure. ECC's tensile strain ranges from 3 to 7% [20–22] and it forms dense cracks with spacing as narrow as 3 to 6 mm post-cracking [23–25]. Additionally, ECC outperforms concrete in shear bearing and deformation resistance [26] and exhibits excellent crack width control, deformability, self-healing ability, and damping characteristics in various structural components [27–30], making it an ideal seismic material.

Zhang et al. [31] studied ECC coupling beams' shear strength and seismic resistance, finding superior performance in these aspects. Suryanto et al. [32] compared the earthquake resistance of external ECC beam–column joints to those without ECC, with ECC joints showing better performance. Cai et al. [33] investigated the ductility, stiffness, and energy dissipation of ECC/RC composite frames compared to RC frames, noting significant improvements in the ECC/RC frames. Khan et al. [34,35] demonstrated that ECC beam–column joints without shear reinforcement still surpassed the code-specified shear resistance. Ye et al. [36] and Yang et al. [37] conducted low cyclic loading experiments on RECC-coupled shear walls, finding that using ECC materials in cast in situ connection areas significantly improved the seismic energy dissipation capacity compared to specimens without ECC. Yang et al. [38] studied the principles of stress and energy dissipation under low cyclic loading through finite element analysis, revealing the failure mechanisms and internal force distribution in prefabricated ECC/RC shear walls. However, to date, no study on the optimization design of prefabricated ECC/RC composite coupled shear walls based on seismic energy dissipation mechanism or comprehensive quantitative analysis has been conducted.

In this paper, the seismic energy dissipation performance of the prefabricated ECC/RC composite coupled shear wall structure will be studied by combining experimental and numerical analysis. The optimal design will be studied based on the seismic energy dissipation mechanism. Through comprehensive numerical quantitative analysis, the optimization design suggestions will be put forward. This study will fill the gap in the field of seismic energy dissipation optimization design of the prefabricated ECC/RC composite coupled shear walls. This work is original and novel and has great scientific and engineering significance.

2. Experimental Investigation

2.1. Specimen Design

For the experimental study, a half-scale, two-story spatial structure representing a prefabricated ECC/RC composite shear wall structure was conceived. This complex assembly comprises eight prefabricated flange walls, eight web walls, four composite coupling beams, eight lateral beams, four bases, and six composite floors. The design is inspired by the original wall system of Zhongnan Century City in Jiangsu Province, China, aligning with the 'Technical specification for precast concrete structures' [39]. In strategic locations, such as the juncture of the coupling beam and the prefabricated concrete wall pier, and atop the composite coupling beams, ECC material is meticulously placed, as depicted in Figure 1. The visual representation (Figure 1) deliberately omits the reinforcement details of the prefabricated elements to maintain clarity. Table 1 enumerates the identification and reinforcement specifics of these components. Dimensions of coupling beams are noted, with a length of 920 mm, whereas the heights of the precast concrete section and the cast-in-place ECC measure 150 mm and 100 mm, respectively, resulting in a span-to-height ratio of 3.68. The beam–wall connection zone dimensions vary across two layers, with the first and second layers measuring 160 mm in width and 450 mm and 350 mm in height, respectively. Vertical integration of the steel bars within the precast walls is achieved through steel sleeves and expansive cement mortar. Meanwhile, horizontal connections—encompassing flange to web wall, beam to wall, beam to floor, and wall to floor—are executed with either cast-in-place concrete or ECC, ensuring structural cohesion.

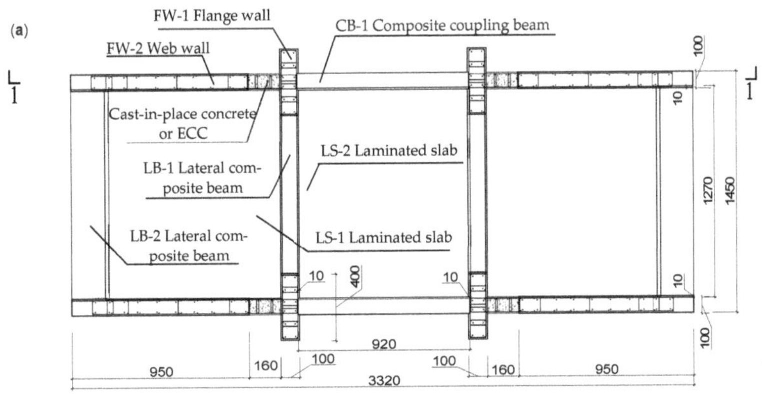

Figure 1. *Cont.*

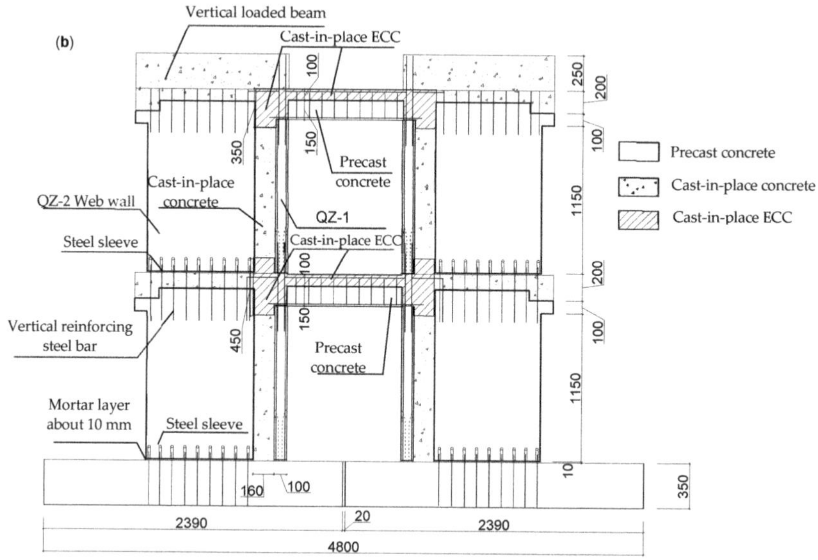

Figure 1. Graphs of the two-story spatial structure specimen of the prefabricated ECC/RC combined shear wall. (**a**) Planar graph; (**b**) 1-1 section graph.

Table 1. The ID and reinforcement details of prefabricated components.

Component	FW-1 Flange Wall	FW-2 Web Wall	CB-1 Composite Coupling Beam	LB-1 Lateral Composite Beam	LB-2 Lateral Composite Beam	LS-1 Laminated Slab	LS-2 Laminated Slab	B-1 Base
Number	8	8	4	4	4	4	2	4
Geometric dimension (mm)	400 × 100 × 1190	850 × 100 × 1340	100 × 150 × 920	100 × 100 × 970	200 × 100 × 1270	945 × 50 × 1270	920 × 50 × 1270	2390 × 900 × 350
Longitudinal reinforcement	12C6	8C6+10A6	3C8	3C12	3C12	A6@100	A6@100	36C4
Horizontal reinforcement	A6@100	A6@100	—	—	—	A6@100	A6@100	C8@50
Stirrups	A4@50	A4@50	C6@50	C6@50	C6@50	—	—	—

Note: C-HRB400 steel bar, A-HPB300 steel bar.

2.2. Material Properties

The examination of material properties reveals distinct compressive strengths for the concrete used in the precast elements and the concrete used in situ, recorded at 36.8 MPa and 32.6 MPa, respectively. Engineered cementitious composites (ECC) exhibit a compressive strength of 35.2 MPa and a tensile strength of 4.6 MPa, alongside an ultimate tensile strain of 3.4%. The reinforcing bars, HPB300 and HRB400, demonstrate yield strengths of 309 MPa and 416 MPa, respectively. The mechanical characteristics of both concrete and ECC, along with the reinforcing steel's properties, are systematically cataloged in Tables 2 and 3.

Table 2. Mechanical properties of concrete and ECC.

Materials	Compressive Strength $f_{cu,k}$ (MPa)	Tensile Strength $f_{t,k}$ (MPa)	Modulus of Elasticity E_s (GPa)	Ultimate Tensile Strain ε_{su}
Precast concrete	36.8	2.2	3.15	—
Cast-in-place concrete	32.6	2.1	3.12	—
ECC	35.2	4.6	1.54	3.4%

Table 3. Mechanical properties of the steel rebars.

Steel Type	Yield Strength f_y (MPa)	Ultimate Strength f_u (MPa)	Modulus of Elasticity E_s (GPa)
HPB300	309	358	209
HRB400	416	489	200

2.3. Testing Methodology

The specimen's foundation is securely fastened to the structural laboratory ground using ground anchor screws passing through pre-designed holes. Annotations 8 and 9 in Figure 2 are the ground anchor screws and laboratory ground, respectively. Vertically, the structure's axial bearing capacity is subjected to a 24% load via eight hydraulic jacks, which tension steel strands. Annotations 4 and 7 marked in Figure 2 are hydraulic jacks and steel strands, respectively. This force is then evenly distributed across the specimen's summit through transfer beams (annotation 3 in Figure 2), as depicted in Figure 2's schematic of the cyclic loading apparatus.

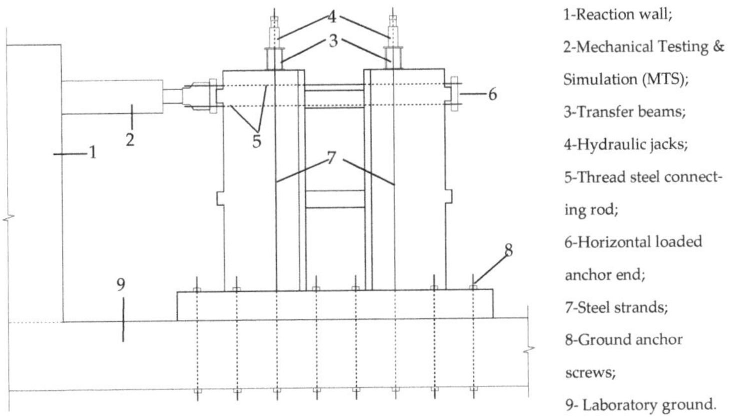

Figure 2. Schematic representation of the cyclic loading device.

Horizontal cyclic loading is administered using a mechanical testing and simulation (MTS) system (annotation 2 in Figure 2), anchored to the reaction wall (annotation 1 in Figure 2) with a maximum force of 1000 kN. The horizontal load is transmitted to the end of the specimen through the threaded steel connecting rods, so that both thrust and tension can be applied, as illustrated in annotations 5 and 6 in Figure 2. The horizontal cyclic loading adopts the slow continuous loading method, and the loading rate is 0.5 mm/s [40]. Initially, in the specimen's elastic phase prior to yielding, loading cycles are managed under load control, and are incremented by 50 kN until the anticipated yield load (P_y) is reached, defined by the yield displacement (Δ_y). After yielding, loading cycles transition to displacement control, progressing through incremental stages of Δ_y, $2\Delta_y$, $3\Delta_y$, ... Each load level undergoes a single cycle, whereas each displacement level is subjected to three cycles. The cyclic loading sequence is detailed in Figure 3. Testing concludes when the load diminishes to 85% of the maximum load, at which point the specimen is deemed to have failed.

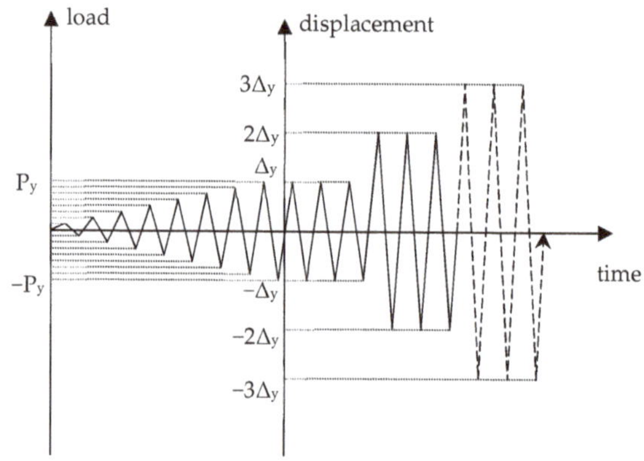

Figure 3. Time history plot of the cyclic loading.

2.4. Damage and Failure Characteristics

The damage and failure of the prefabricated ECC/RC composite shear wall structure are mainly manifested in the cracking of the bottom of the shear walls and the coupling beams, which can be seen in Figure 4. The cracks in the web walls are horizontal cracks at the initial stage of loading, and gradually develop along the inclined direction with the increase in horizontal load, and finally show as bending-shear cracks. The cracks in the flange walls are horizontal cracks, which eventually show bending failure. The cracking of the coupling beams is mainly concentrated at the ends and the cast-in-place ECC area, and there are many fine cracks in the ECC.

(a)

Figure 4. *Cont.*

(**b**) (**c**)

Figure 4. Damage and failure of the prefabricated ECC/RC composite shear wall structure. (**a**) Integral structure; (**b**) web wall; (**c**) flange wall.

2.5. Load–Displacement Hysteresis Behavior

Figure 5 illustrates the load–displacement hysteresis loops, which are characterized by a bow shape, highlighting the specimen's robust energy dissipation capability and minimal residual deformation. The load–displacement skeleton curve, detailed in Figure 6, delineates critical stages, such as the cracking, yielding, peak, and failure points. Initial crack formation was observed at a displacement of 3.64 mm under a horizontal force of 248.7 kN. Yielding occurred when the displacement reached 7.84 mm with an applied load of 524 kN. The specimen achieved its maximum load capacity at 991.7 kN, corresponding to a displacement of 33.67 mm. The failure threshold was identified when the load diminished to 836 kN and displacement extended to 50.5 mm. Table 4 compiles the mechanical properties extracted from these observations.

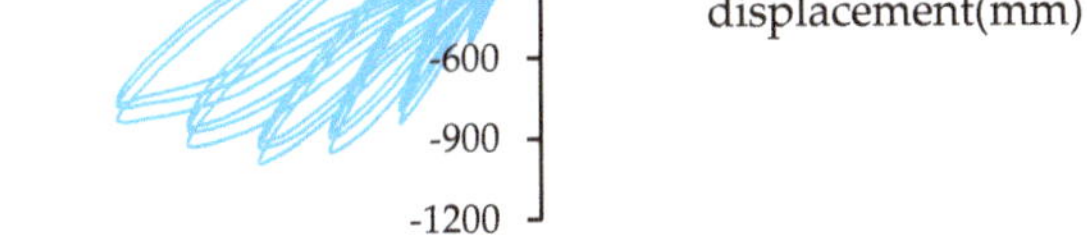

Figure 5. Load–displacement hysteresis curves.

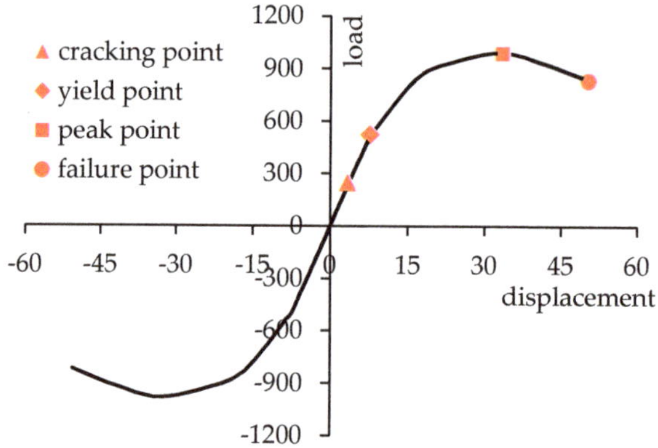

Figure 6. Load–displacement skeleton curve.

Table 4. Mechanical characteristic values of the specimen.

Mechanical Properties	Cracking Point	Yield Point	Peak Point	Failure Point
Displacement (mm)	3.64	7.84	33.67	50.5
Load (kN)	248.7	524.0	991.7	836.0

3. Numerical Simulation

In this study, the finite element analysis software ABAQUS 2020, renowned for its comprehensive capabilities, serves as the foundation for the numerical simulations. The process of developing and analyzing models encompasses several critical steps outlined below.

3.1. Material Constitutive Relations

The simulation's framework integrates the behavior of concrete, ECC, and steel reinforcement. The uniaxial stress–strain curve for concrete is adopted from the "Code for Design of Concrete Structures" (2015 Edition) (GB50010-2010) [41], ensuring compliance with established standards. The ECC's stress–strain profile is derived from the methodology introduced by the research team led by Pan [42], highlighting the innovative application of ECC. Furthermore, the simulation incorporates damage factors for both concrete and ECC, following the guidelines proposed by Zhang Jin and colleagues [43]. For the modeling of steel reinforcement, a bilinear stress–strain curve is utilized, providing a detailed representation, as depicted in Figure 7.

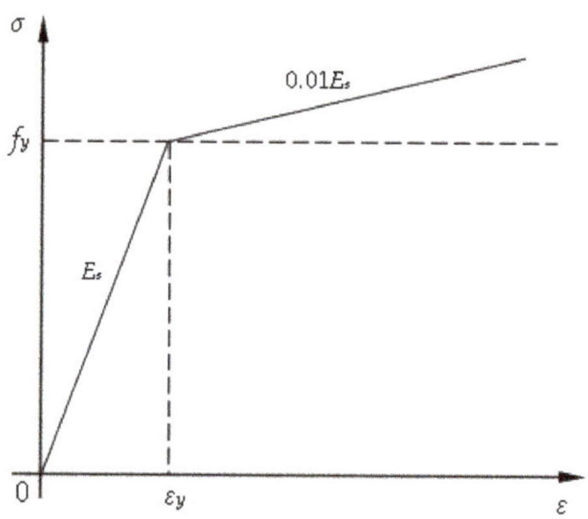

Figure 7. Stress–strain curve of the steel bars.

3.2. Element Type and Model Building

In the realm of finite element analysis, both the concrete and ECC elements are depicted using eight-node hexahedral linear reduced-integration elements (C3D8R) [42]. The prefabricated joints' vertical reinforcements are modeled with two-node spatial linear beam elements (B31), and the additional reinforcement utilizes space truss elements (T3D2).

Interactions between concrete surfaces and between concrete and ECC are modeled via surface-to-surface contact. For normal direction interactions, a hard contact strategy is adopted, permitting potential separation, while tangential movements are governed by a penalty friction model with a friction coefficient of 0.4 [44].

The simulation constructs a half-scale, two-story spatial structure of a prefabricated ECC/RC combined shear wall, reflective of the experimental specimen's dimensions. The finite element model's depiction, including the application of ECC at critical junctions, such as beam-to-wall and within the upper extents of composite coupling beams, is presented in Figure 8a. Figure 8b showcases the steel reinforcement layout within the model. Given the negligible slip between the steel reinforcement and concrete, as well as between the steel and ECC under applied forces, the model assumes effective bonding, employing the "embedded region" contact feature for interface adherence in ABAQUS.

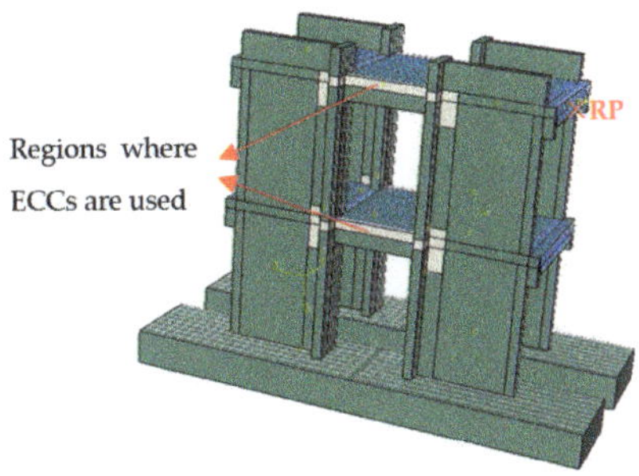

(a)

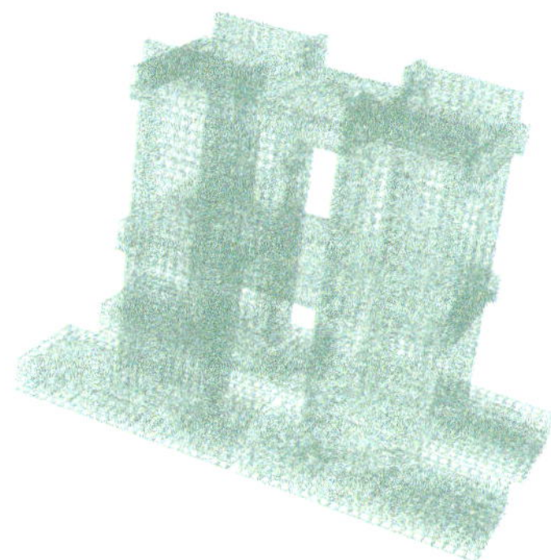

(b)

Figure 8. Finite element analysis model of the prefabricated ECC/RC combined shear wall. (**a**) Finite element model; (**b**) steel cage of the spatial structure specimen model.

3.3. Boundary Conditions and Loading Methods

The foundation of the modeled structure is securely anchored at its base, mirroring the experimental conditions. Vertical forces are consistently applied across the loading beam, as shown in Figure 9a. Horizontally, the model undergoes reversed cyclic loading, initiated through force-displacement techniques. Initially, force-based loading is adopted, proceeding in stages with each increment defined as a 50 kN multiple. Following the attainment of yield displacement, the procedure transitions to displacement-based loading, employing increments equal to integer multiples of the yield displacement. This phase involves conducting three cycles at each level until structural failure is observed. The loading sequence reflects the experimental approach, detailed in Figure 3. A reference point (RP) is designated at the spatial structure model's apex, linked to nodes at both extremities of the loading beam, enabling the implementation of horizontal displacements at RP, as depicted in Figure 9b.

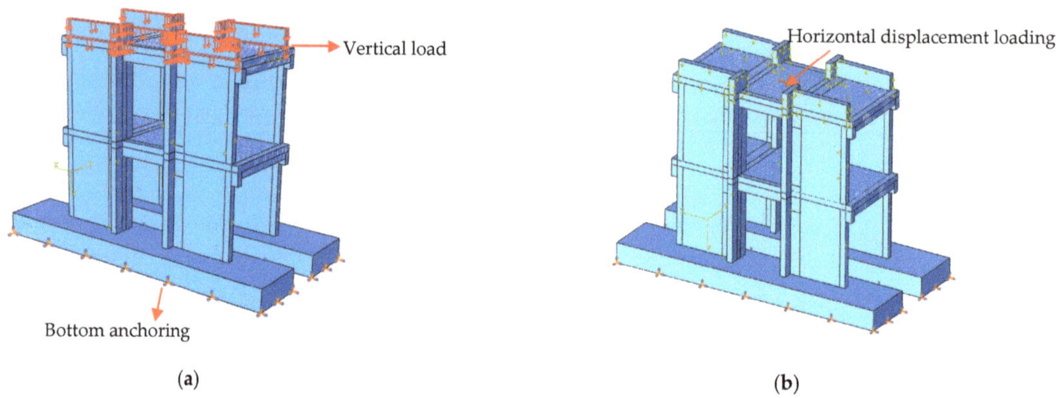

Figure 9. Boundary conditions and loading methods. (**a**) Boundary conditions and vertical loads; (**b**) low-cycle repeated horizontal displacement loading.

3.4. Mesh Subdivision

Utilizing the swept meshing approach guarantees consistency in mesh generation, with a specified cutting surface to ensure uniformity across all elements. This includes steel reinforcements, wall segments, coupling beams, floor sections, loading beams, and foundational elements, each with a mesh dimension uniformly established at 0.05 m. The spatial model's mesh configuration is illustrated in Figure 10.

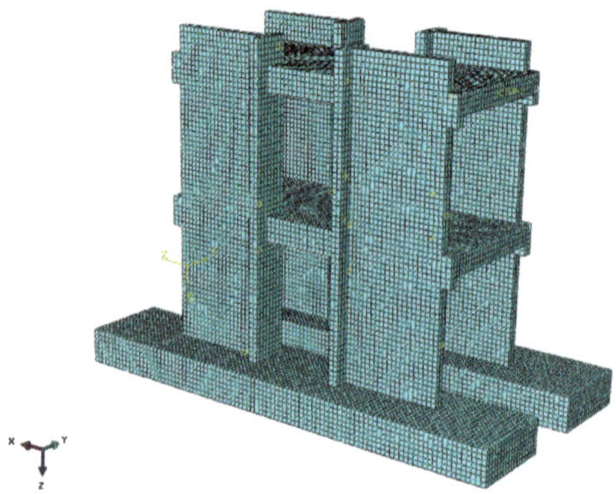

Figure 10. Finite element model meshing.

3.5. Analytical Calculations

The computational framework is segmented into two distinct phases. Initially, the foundation is immobilized to define inter-component contact dynamics, paving the way for vertical force application. The subsequent phase involves constraining lateral movements parallel to the wall's plane on the loading beam's sides, ensuring structural equilibrium under load. Horizontal forces are then engaged. For iterative computation, the Newton–Raphson technique is applied, with step size adjustments to address convergence or efficiency challenges.

3.6. Damage and Crack Distribution

The damage and crack distribution of the concrete and ECC in ABAQUS is reflected by the tensile damage factor. The closer the damage factor is to 1.0, the more serious the damage to the concrete or ECC is. Figure 11 shows the distribution and variation of the tensile damage factor of the spatial structure specimen of the prefabricated ECC/RC combined shear wall structure.

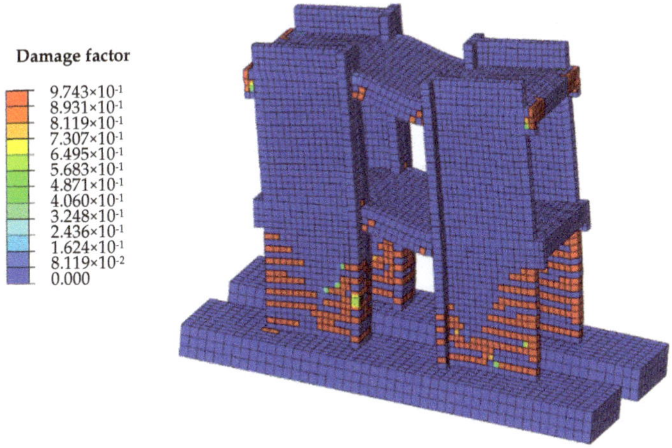

Figure 11. Damage and crack distribution of the prefabricated ECC/RC composite shear wall structure.

3.7. Comparison of Numerical and Experimental Results

In order to make the comparison between the test and the calculation results clear, the web wall and the flange wall are intercepted from the integral structure for comparison. Figure 12 is the comparison of experimentally measured cracks and finite element simulation cracks. Figure 12 shows that the webs are mainly flexural shear diagonal cracks, and the flanges are mainly horizontal bending cracks. The crack pattern simulated by the finite element model is approximately consistent with the actual crack pattern.

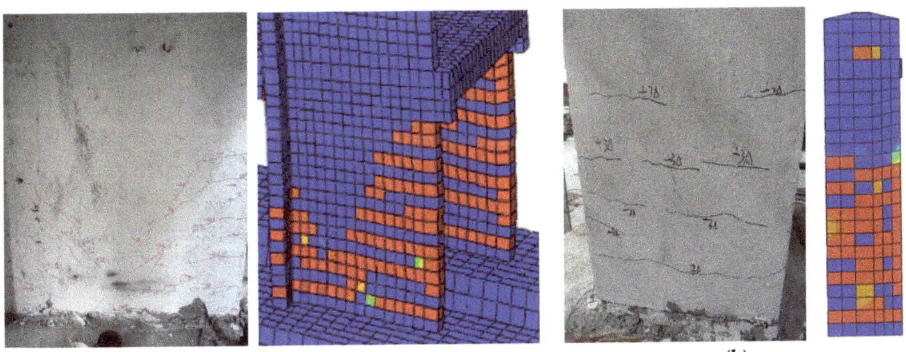

(**a**) (**b**)

Figure 12. Comparison of experimentally measured cracks and finite element simulation cracks. (**a**) Web wall; (**b**) flange wall.

Figures 13 and 14 showcase that the load–displacement hysteresis and skeleton curves from both the simulation and experimental protocols are approximately consistent with the experimental test results. However, the finite element method shows a slightly stiffer rigidity and a lower displacement. One reason is that the finite element calculation method uses the finite degree of freedom to approximate the infinite degree of freedom in the real structure. The finite element solution is equivalent to making the structure deform according to the given shape function. Compared with the real situation, the constraints are increased in the finite element simulation, so the stiffness is larger than the real solution. Correspondingly, the displacement is a little smaller. The second possible reason is that the threaded steel connecting rods that transmit the horizontal load may produce a small irreversible tensile deformation in the later stage of loading, resulting in a length slightly larger than the original length, and the measured displacement of the specimen also increases accordingly. There are two ways to reduce this gap. One is to divide the meshes more finely in the finite element calculation, so that the gap is closer to the infinite degree of freedom, and, thus, closer to the real situation. Another method is to use almost non-deformed steel to make the threaded steel connecting rods to reduce irreversible tensile deformation.

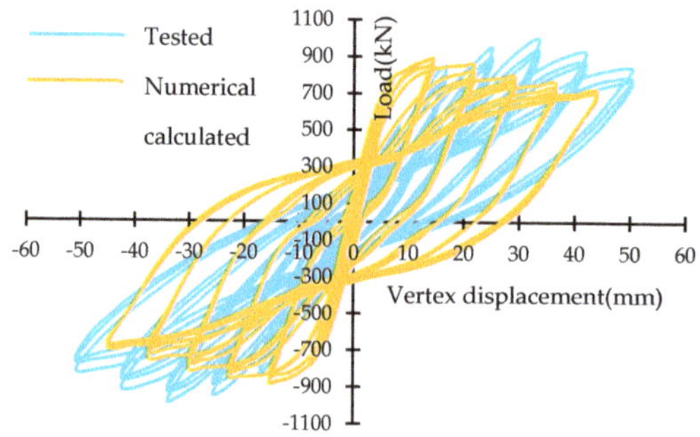

Figure 13. Load–displacement hysteresis curves.

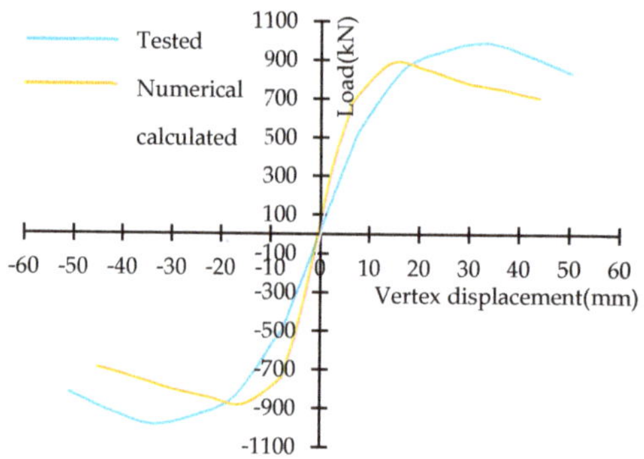

Figure 14. Load–displacement skeleton curves.

Although there are slight differences, the comparison between the experimental and numerical simulation results of the cracking morphology and the load–displacement curves verifies the correctness of the numerical simulation method.

4. Quantitative Analysis of Influence Factors

This section delves into the paramount variables that shape the seismic energy absorption capabilities of such structures. These include the deployment areas of ECC within composite coupling beams and shear walls, ECC's compressive strength, the coupling beams' stirrup ratio, and the tensile strength of longitudinal reinforcements. An exhaustive quantitative examination was undertaken to evaluate the impact of these parameters on seismic energy dissipation metrics, such as load–displacement hysteresis behaviors and energy absorption efficiency, utilizing our refined numerical model. Herein, a comprehensive analysis of these parameters' effects on the structural seismic energy dissipation efficiency is articulated.

4.1. Use Regions of ECC in Composite Coupling Beams

4.1.1. Selection of Parameters

Given its enhanced complexity and cost over traditional concrete, ECC's economic viability, especially in fully incorporating it within coupling beams, warrants scrutiny. This investigation evaluates the seismic energy absorption efficiency of ECC by comparing three coupling beam configurations: fully concrete (structure 1-1), a hybrid of ECC and concrete (structure 1-2), and exclusively ECC (structure 1-3). Consistent across these variants are other factors, like the stirrup ratio, the nature of longitudinal reinforcement, concrete strength, and ECC type, among others. Table 5 enumerates the specific parameters for this analysis, adhering to Chinese standards [40], where HRB400 denotes a 400 MPa tensile strength of the reinforcing bar, and E40 and C40 represent the compressive strengths of ECC and concrete, respectively, each at 40 MPa.

Table 5. Variation parameters of ECC in coupling beams.

Classification of Specimens	Application Regions of ECC in Coupling Beams	Stirrup Ratio of Coupling Beams	Type of Longitudinal Reinforcement	Type of ECC	Type of Concrete
Structure 1-1	None (Concrete coupling beam)	1.13%	HRB400	E40	C40
Structure 1-2	Cast-in-place areas (ECC/RC coupling beams)				
Structure 1-3	Full ECC (ECC coupling beams)				

4.1.2. Load–Displacement Hysteresis Behavior

Figure 15 illustrates the load–displacement hysteresis curves for specimens subjected to low cyclic loading, highlighting the differential impact of ECC application within the coupling beams. Integrating ECC not only postpones the yielding of these beams but also enhances the overall shear wall structure's delay in yielding, thereby augmenting the energy dissipation potential of the structure. The depicted curves showcase a pronounced bow shape. Notably, the pinching effect—a measure of energy dissipation efficiency—is most evident in the purely concrete structure (structure 1-1), diminishing progressively in the ECC-integrated structures (structure 1-2 and structure 1-3), with structure 1-3 exhibiting the least pinching. This indicates that structure 1-3's hysteresis loop is the most complete, signifying superior energy dissipation capability.

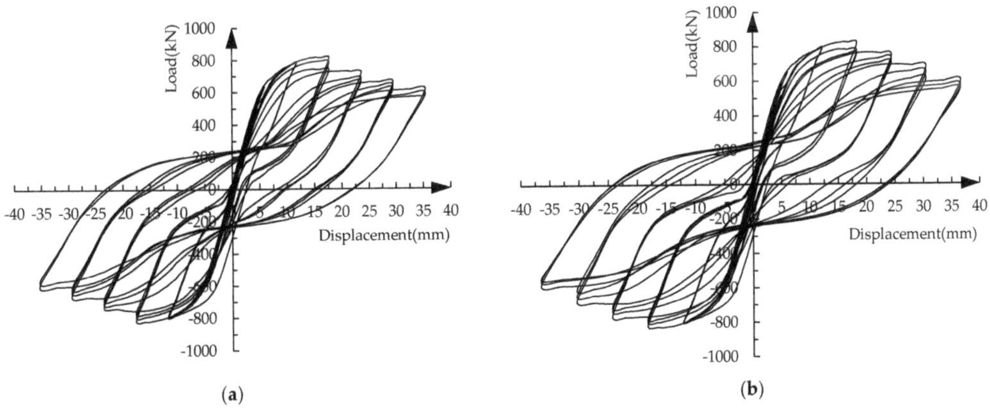

Figure 15. *Cont.*

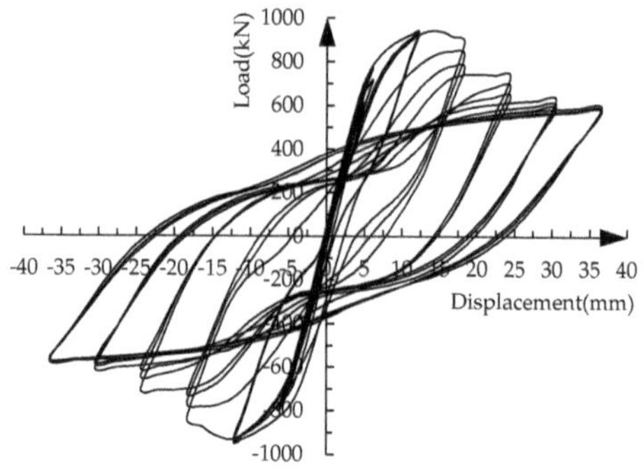

(c)

Figure 15. Load–displacement hysteresis curves. (**a**) Structure 1-1; (**b**) structure 1-2; (**c**) structure 1-3.

4.1.3. Energy Dissipation Capacity

The capability of energy absorption, quantified by the hysteresis loop's enclosed area, is determined through an approximate integral technique. Table 6 presents both the single-cycle energy dissipation metrics and the aggregate energy dissipation figures for the specimens, reflecting different extents of ECC integration within the composite coupling beams. The symbol Δ in Table 6 denotes the yield displacement.

Table 6. Single-cycle/cumulative energy dissipation values (J).

Displacement Loading Level	Number of Cycles	Structure 1-1		Structure 1-2		Structure 1-3	
		Single-Cycle	Cumulative	Single-Cycle	Cumulative	Single-Cycle	Cumulative
1Δ	1	918.5	918.5	1186.2	1186.2	1287.7	1287.7
	2	555.6	1474.1	608.7	1794.9	658.6	1946.3
	3	263.7	1737.8	220.6	2015.5	237.5	2183.8
2Δ	1	6521.6	8259.4	7417.9	9433.4	8062.9	10,246.7
	2	4307.9	12,567.3	5292.6	14,726.0	5752.8	15,999.5
	3	4067.5	16,634.8	5050.2	19,776.2	5489.4	21,488.9
3Δ	1	12,299.2	28,934	17,250.3	37,026.5	18,750.3	40,239.2
	2	12,178.7	41,021.9	14,711.9	51,738.4	15,991.2	56,230.4
	3	12,087.9	53,200.6	12,974.0	64,712.4	14,102.2	70,332.6
4Δ	1	21,063.8	74,264.4	21,897.7	86,610.1	23,801.8	94,134.4
	2	18,255.7	92,520.1	19,782.7	106,392.8	21,502.9	115,155.6
	3	16,788.5	109,308.6	19,339.5	125,732.3	21,021.2	136,658.5
5Δ	1	25,502.7	134,811.3	28,540.1	154,272.4	31,021.9	167,680.4
	2	22,941.6	157,752.9	27,119.2	181,391.6	29,477.4	197,157.8
	3	21,861.5	179,614.4	26,669.4	208,061.0	28,988.5	226,146.3
6Δ	1	30,998.5	210,612.9	37,044.6	244,032.7	40,265.9	266,412.2
	2	29,796.7	240,409.6	35,971.7	281,077.3	39,099.7	305,511.9
	3	28,689.7	269,099.3	35,280.9	316,358.2	38,348.8	343,860.7

The analysis distinctly showcases that integrating ECC, even solely within the cast-in-place sections of coupling beams, significantly bolsters the structure's seismic energy absorption capabilities. As delineated in Table 6, a definitive hierarchy emerges in energy dissipation efficiency: Structure 1-3 surpasses structure 1-2, which in turn exceeds structure 1-1, across both single-cycle and aggregated energy dissipation metrics. This progression underscores an enhancement in energy absorption capabilities correlating with the augmented incorporation of ECC within the coupling beams.

4.2. Use Regions of ECC in Shear Walls

4.2.1. Selection of Parameters

Acknowledging the susceptibility of shear wall lower sections to seismic forces, which frequently result in plastic hinge formation, this segment investigates the impact of varying ECC application extents at the shear wall base to ascertain optimal implementation. Structure 2-1, constructed purely of concrete, contrasts with structure 2-2 through structure 2-5, which progressively integrate ECC from 200 mm to 800 mm at the base, corresponding to incremental structural height percentages. Consistency is maintained across other variables, such as the stirrup ratio in coupling beams and the specifications of longitudinal reinforcement and concrete types. Detailed parameters for this analysis are encapsulated in Table 7.

Table 7. Variation parameters of ECC in shear walls.

Classification of Specimens	ECC Application at the Base of Shear Walls	Stirrup Ratio of Coupling Beams	Type of Longitudinal Reinforcement	Type of ECC	Type of Concrete
Structure 2-1	0	1.13%	HRB400	E40	C40
Structure 2-2	7% structural height (200 mm)				
Structure 2-3	14% structural height (400 mm)				
Structure 2-4	21% structural height (600 mm)				
Structure 2-5	28% structural height (800 mm)				

4.2.2. Load–Displacement Hysteresis Behavior

Illustrated in Figure 16 are the load–displacement hysteresis curves for specimens subjected to low cyclic loading. These curves demonstrate a more pronounced completeness as the application height of ECC at the base of the shear walls increases, signifying a notable improvement in the structure's capacity for energy dissipation.

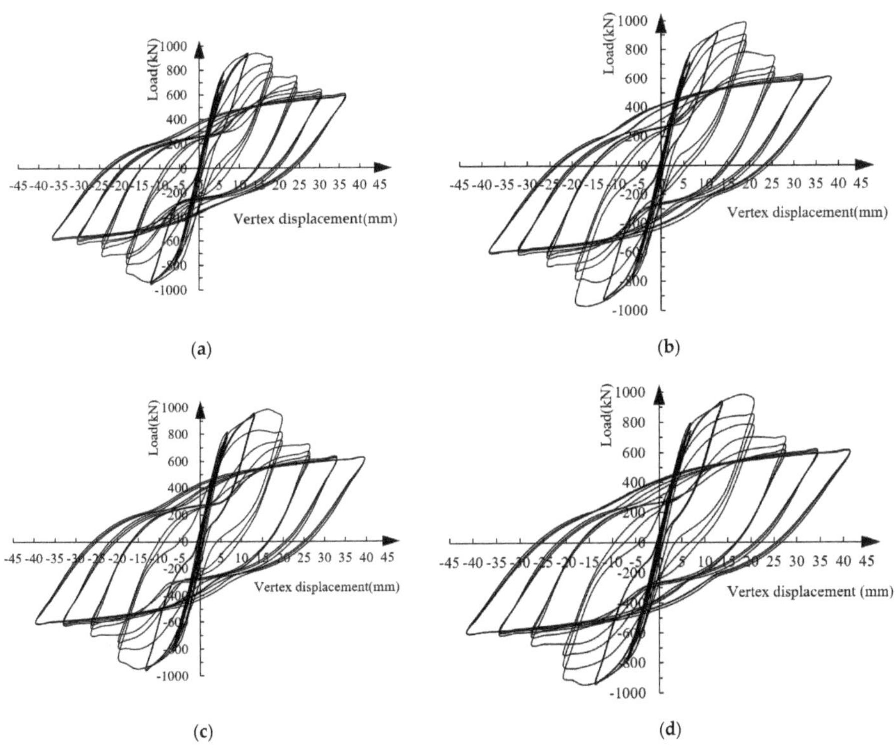

Figure 16. *Cont.*

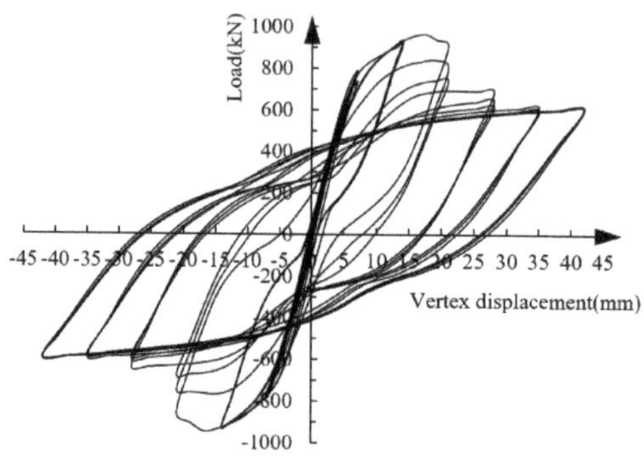

(e)

Figure 16. Load–displacement hysteresis curves. (**a**) Structure 2-1; (**b**) structure 2-2; (**c**) structure 2-3; (**d**) structure 2-4; (**e**) structure 2-5.

4.2.3. Energy Dissipation Capacity

Table 8 details the energy dissipation metrics for five structural configurations across both single cycles and cumulative impacts. Initially, Structure 2-1 outperforms structure 2-2 in single-cycle energy dissipation, a phenomenon highlighted in Table 8. This stage, marking the specimens' initial yielding phase, reveals the minimal yet significant influence of ECC application. At this juncture, the inherent stiffness of the structural elements predominantly governs energy dissipation, with the rigidity of concrete contributing more significantly than that of ECC, thus favoring structure 2-1's energy dissipation profile over structure 2-2.

With the elevation of ECC incorporation to 400 mm, an increase in yield points is observed, overshadowing the stiffening effect of concrete, and setting a new comparison where structure 2-2 is surpassed by structure 2-3. Incrementing ECC's presence to between 600 and 800 mm at the shear wall's base diminishes the yield displacement's role in energy dissipation, reshaping the hierarchy to structure 2-3 outperforming structure 2-4 and structure 2-5. Moreover, extending displacement loading to double the yield displacement reconfigures the relationship across single-cycle and cumulative energy dissipation figures, establishing a trend where greater ECC implementation in shear walls correlates with heightened energy dissipation capacity, evidenced by the no ECC to 800 mm increments.

Table 8. Single cycle/cumulative energy dissipation of the five structural specimens (J).

Displacement Loading Level	Number of Cycles	Structure 2-1		Structure 2-2		Structure 2-3		Structure 2-4		Structure 2-5	
		Single-Cycle	Cumulative	Single-Cycle	Cumulative	Single-Cycle	Cumulative	Single-Cycle	Cumulative	Single-Cycle	Cumulative
1Δ	1	1193.5	1193.5	1169.7	1169.7	1715.6	1715.6	1491.8	1491.8	1453.9	1453.9
	2	647.7	1841.2	502.3	1672	768.3	2483.9	545.5	2037.3	515	1968.9
	3	216.2	2057.4	297.6	1969.6	412.3	2896.2	308.6	2345.9	332	2300.9
2Δ	1	7975.3	10,032.7	7652.6	9622.2	8556.3	11,452.5	8651.8	10,997.7	8666.9	10,967.8
	2	5674.6	15,707.3	5863.1	15,485.3	7367.4	18,819.9	7160.8	18,158.5	7238.6	18,206.4
	3	5397.3	21,104.6	5771.2	21,256.5	7005.4	25,825.3	6922.4	25,080.9	7012.2	25,218.6
3Δ	1	18,649.2	39,753.8	18,344.2	39,600.7	22,023.7	47,849	21,322.3	46,403.2	22,419	47,637.6
	2	15,112.3	54,866.1	17,998.3	57,599	20,969.8	68,818.8	20,720	67,123.2	21,431.9	69,069.5
	3	14,073.5	68,939.6	15,419.6	73,018.6	17,581.2	86,400	19,805	86,928.2	19,138.9	88,208.4
4Δ	1	23,429.7	92,369.3	26,425.3	99,443.9	27,936.5	114,336.5	29,570.9	116,499.1	29,906.4	118,114.8
	2	21,113.2	113,482.5	24,323.6	123,767.5	25,400.1	139,736.6	27,143.4	143,642.5	27,295.6	145,410.4
	3	20,786.3	134,268.8	23,710.5	147,478	24,652.7	164,389.3	25,575.5	169,218	27,270.4	172,680.8
5Δ	1	30,786.5	165,055.3	33,976.2	181,454.2	36,745	201,134.3	37,849.2	207,067.2	38,685.8	211,366.6
	2	29,011.6	194,066.9	33,178.8	214,633	36,516.6	237,650.9	37,338.2	244,405.4	37,798.4	249,165
	3	28,653.2	222,720.1	33,074.3	247,707.3	35,751	273,401.9	36,303	280,708.4	36,649.8	285,814.8
6Δ	1	39,862.9	262,583	44,226.3	291,933.6	47,475.5	320,877.4	48,948.2	329,656.6	48,656.9	334,471.7
	2	38,865.2	301,448.2	42,983.9	334,917.5	45,928.6	366,806	47,494.3	377,150.9	47,598.2	382,069.9
	3	38,012.3	339,460.5	42,656.2	377,573.7	45,399.4	412,205.4	46,998.2	424,149.1	47,196.5	429,266.4

Table 8 also illustrates the cumulative energy dissipation values for structures 2-1 to 2-5, which are 339,460.5 J, 377,573.7 J, 412,205.4 J, 424,149.1 J, and 429,266.4 J, respectively, after undergoing the third cycle of 6Δ loading displacement. When compared to structure 2-1, the cumulative energy dissipation of structures 2-2 to 2-5 shows an increase of 11.2%, 21.4%, 24.9%, and 26.5%, respectively. Notably, the application of ECC material at the bottom 400 mm of the shear wall results in a cumulative energy dissipation increase of over 20% compared to the specimen without ECC. As ECC usage at the bottom of the shear wall continues to rise, there is a corresponding increase in cumulative energy dissipation, albeit at a diminishing rate. For instance, the energy dissipation of structure 2-3 is 9.2% higher than that of structure 2-2, while the increase from structure 2-2 to structure 2-1 is more modest at 2.9%.

Optimal energy dissipation performance is achieved with the incorporation of ECC at 400 mm from the base of the shear wall. The distinctive properties of ECC, including its enhanced ductility, damping capabilities, and superior resistance to deformation under tensile, compressive, and shear forces—despite its comparative lower stiffness to traditional concrete—play a pivotal role in dictating the overall energy dissipation efficiency. This efficiency is critically influenced by the material's deformation capacity, stiffness, and load-bearing potential. Consequently, it is advised that the ideal proportion of ECC to be utilized in precast shear walls is established at 14% (400 mm) of the total structural height.

4.3. *Strength of ECC*

4.3.1. Selection of Parameters

Parameter analysis from Sections 4.1 and 4.2 indicate optimal seismic performance when ECC is utilized in specific areas, namely the coupling beams and the lower 400 mm of shear walls. We varied the ECC strength in these regions to assess its impact, using the ECC types E40, E60, and E80, corresponding to compressive strengths of 40 MPa, 60 MPa, and 80 MPa, respectively. Other unspecified parameters are the same. The variation parameters of the ECC strength are shown in Table 9.

Table 9. The variation parameters of the ECC strength.

Classification of Specimens	Application Areas of ECC	Stirrup Ratio of Coupling Beams	Type of Longitudinal Reinforcement	Type of ECC	Type of Concrete
Structure 3-1 Structure 3-2 Structure 3-3	Coupling beams and the lower 400 mm of shear walls	1.13%	HRB400	E40 E60 E80	C40

4.3.2. Load–Displacement Hysteresis Behavior

Figure 17 illustrates that the hysteresis behavior of the specimens is relatively unaffected by ECC strength variations, indicating that ECC's energy dissipation capacity is not significantly influenced by its compressive strength.

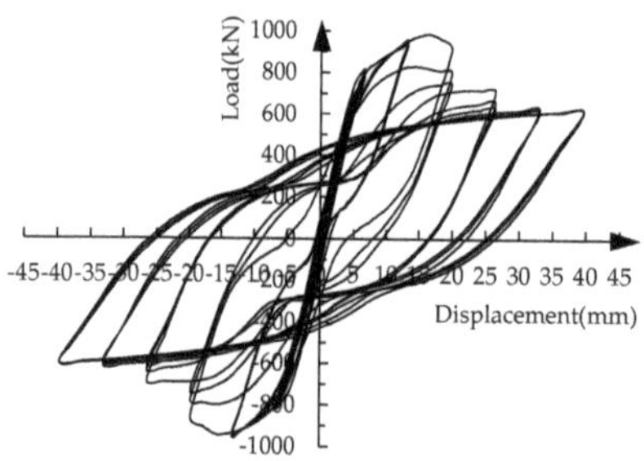

(a)

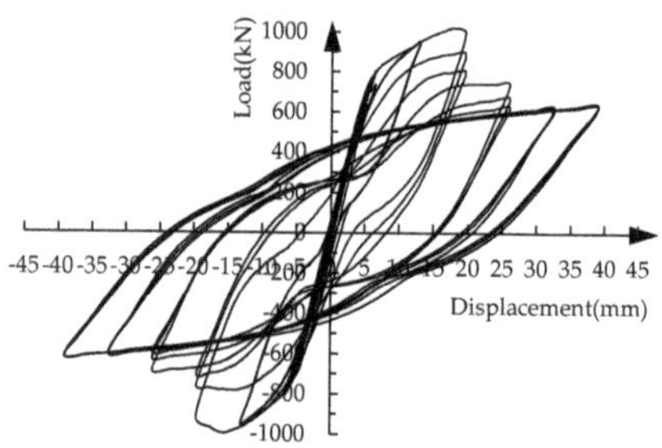

(b)

Figure 17. *Cont.*

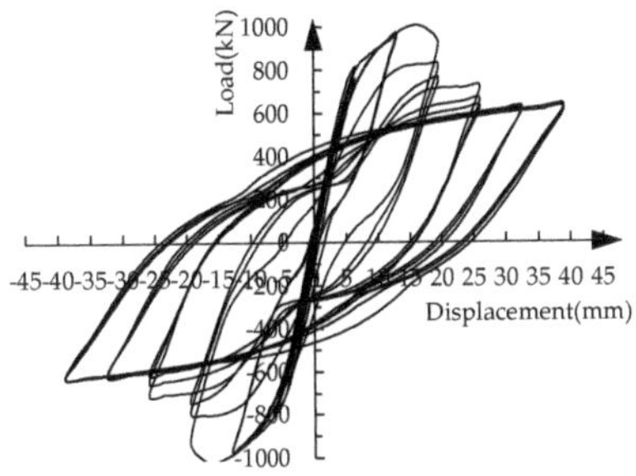

(c)

Figure 17. Load–displacement hysteresis curves. (**a**) Structure 3-1; (**b**) structure 3-2; (**c**) structure 3-3.

4.3.3. Energy Dissipation Capacity

Table 10 presents the energy dissipation data under cyclic loading. Initially, higher strength ECC shows greater energy dissipation, likely due to increased stiffness. However, as loading progresses, the trend reverses, with lower strength ECC demonstrating superior energy dissipation. This highlights the importance of balancing strength and ductility in ECC for optimal seismic performance.

Table 10. Single-cycle/cumulative energy dissipation with different ECC strengths (J).

Displacement Loading Level	Number of Cycles	Structure 2-1		Structure 2-2		Structure 2-3	
		Single-Cycle	Cumulative	Single-Cycle	Cumulative	Single-Cycle	Cumulative
1Δ	1	1797.9	1797.9	1827.7	1827.7	1844.8	1844.8
	2	851.8	2649.7	897.2	2724.9	921.5	2766.3
	3	496.2	3145.9	539.8	3264.7	566.3	3332.6
2Δ	1	8679.7	11,825.6	7850.5	11,115.2	8251.1	11,583.7
	2	7490.8	19,316.4	4902.7	16,017.9	5061.7	16,645.4
	3	7128.8	26,445.2	4347.2	20,365.1	4465.5	21,110.9
3Δ	1	22,147.1	48,592.3	20,359.9	40,725	20,648.2	41,759.1
	2	21,093.2	69,685.5	19,888	60,613	19,996.1	61,755.2
	3	17,704.6	87,390.1	16,238.3	76,851.3	16,635	78,390.2
4Δ	1	28,071	115,461.1	26,369	103,220.3	26,744.5	105,134.7
	2	25,534.6	140,995.7	23,050.5	126,270.8	23,979.7	129,114.4
	3	24,787.2	165,782.9	22,108.2	148,379	22,482.1	151,596.5
5Δ	1	36,879.5	202,662.4	35,549.4	183,928.4	35,664.1	187,260.6
	2	36,651.1	239,313.5	32,085.4	216,013.8	32,158.9	219,419.5
	3	35,885.5	275,199	30,333.9	246,347.7	30,159.7	249,579.2
6Δ	1	47,613.3	322,812.3	45,413.8	291,761.5	46,495.3	296,074.5
	2	46,066.4	368,878.7	43,862.2	335,623.7	44,070.5	340,145
	3	45,537.2	414,415.9	42,948.7	378,572.4	42,525.1	382,670.1

4.4. Stirrup Ratio of Coupling Beams

4.4.1. Selection of Parameters

This study examines the impact of varying stirrup ratios in coupling beams. The coupling stirrup ratios of structure 4-1, structure 4-1, and structure 4-1 are 0.50%, 1.13%, and 2.01% (corresponding to diameters of 4 mm, 6 mm, and 8 mm), respectively. The configurations of the structural specimens are detailed in Table 11.

Table 11. Configuration parameters of structural specimens.

Classification of Specimens	Application Areas of ECC	Stirrup Ratio of Coupling Beams	Type of Longitudinal Reinforcement	Type of ECC	Type of Concrete
Structure 4-1	Coupling beams and the lower 400 mm of shear walls	0.50%	HRB400	E40	C40
Structure 4-2		1.13%			
Structure 4-3		2.01%			

4.4.2. Load–Displacement Hysteresis Behavior

Figure 18 presents the hysteresis curves for the specimens. The similarity across different stirrup ratios suggests a negligible effect on the load–displacement behavior within the parameters of this study.

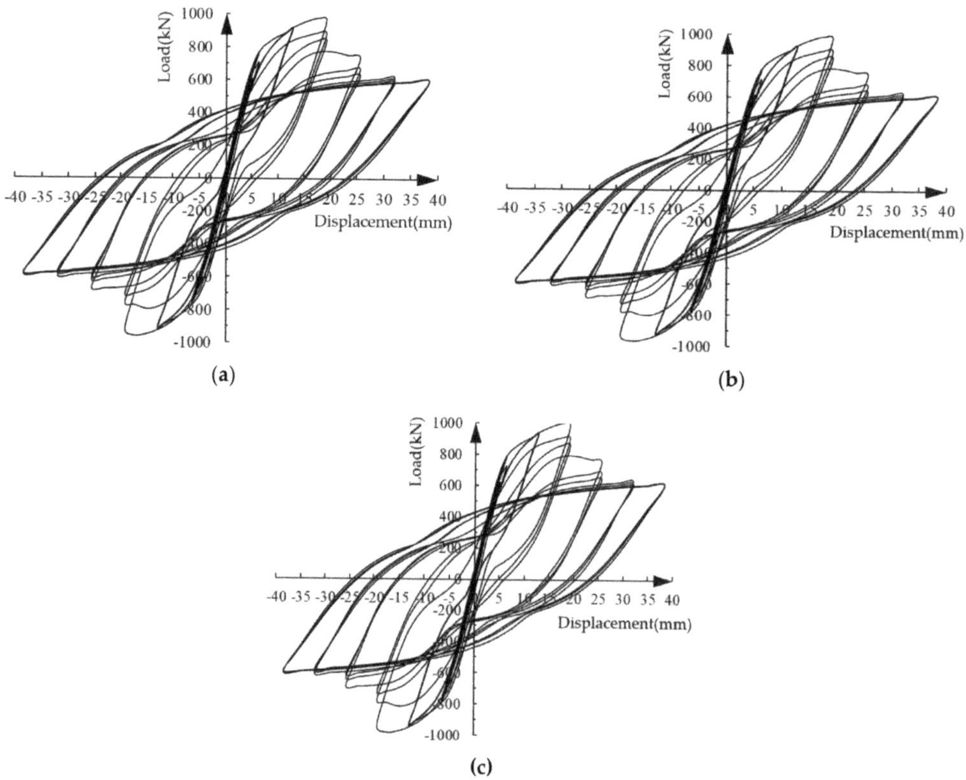

Figure 18. Load–displacement hysteresis curves with varied stirrup ratios. (**a**) Structure 4-1; (**b**) structure 4-2; (**c**) structure 4-3.

4.4.3. Energy Dissipation Capacity

Table 12 compares the energy dissipation values across different stirrup ratios. The data reveals only slight differences in both single-cycle and cumulative energy dissipation values, suggesting a minimal impact of stirrup ratio variations on energy dissipation.

Table 12. Single-cycle/cumulative energy dissipation with different stirrup ratios of coupling beams (J).

Displacement Loading Level	Number of Cycles	Structure 4-1		Structure 4-2		Structure 4-3	
		Single-Cycle	Cumulative	Single-Cycle	Cumulative	Single-Cycle	Cumulative
1Δ	1	1791.2	1791.2	1797.9	1797.9	1822.1	1822.1
	2	846.3	2637.5	851.8	2649.7	861.6	2683.7
	3	494.2	3131.7	496.2	3145.9	502.4	3186.1
2Δ	1	8592.9	11,724.6	8679.7	11,825.6	8809.9	11,996
	2	7415.9	19,140.5	7490.8	19,316.4	7603.2	19,599.2
	3	7057.5	26,198	7128.8	26,445.2	7235.7	26,834.9
3Δ	1	21,925.6	48,123.6	22,147.1	48,592.3	22,479.3	49,314.2
	2	20,882.3	69,005.9	21,093.2	69,685.5	21,409.6	70,723.8
	3	17,527.6	86,533.5	17,704.6	87,390.1	17,970.2	88,694
4Δ	1	27,790.3	114,323.8	28,071	115,461.1	28,492.1	117,186.1
	2	25,279.3	139,603.1	25,534.6	140,995.7	25,917.6	143,103.7
	3	24,539.3	164,142.4	24,787.2	165,782.9	25,159.0	168,262.7
5Δ	1	36,510.7	200,653.1	36,879.5	202,662.4	37,432.7	205,695.4
	2	36,284.6	236,937.7	36651.1	239,313.5	37,200.9	242,896.3
	3	35,526.6	272,464.3	35,885.5	275,199	36,423.8	279,320.1
6Δ	1	47,137.2	319,601.5	47,613.3	322,812.3	48,327.5	327,647.6
	2	45,605.7	365,207.2	46,066.4	368,878.7	46,757.4	374,405
	3	45,081.8	410,289	45,537.2	414,415.9	46,220.3	420,625.3

4.5. Strength of Longitudinal Reinforcement

4.5.1. Selection of Parameters

This section evaluates the impact of varying longitudinal reinforcement strengths on structural performance. The longitudinal reinforcements used in structure 5-1, structure 5-2, and structure 5-3 are HRB335, HRB400, and HRB500, respectively. The ECC usage range is consistent with Section 4.4. Table 13 details the parameter configurations for the specimens.

Table 13. Configuration parameters of structural specimens.

Classification of Specimens	Application Areas of ECC	Stirrup Ratio of Coupling Beams	Type of Longitudinal Reinforcement	Type of Concrete	Type of ECC
Structure 5-1	Coupling beams and the lower 400 mm of shear walls	1.13%	HRB335	C40	E40
Structure 5-2			HRB400		
Structure 5-3			HRB500		

4.5.2. Load–Displacement Hysteresis Behavior

Figure 19 presents the hysteresis curves for the specimens. An increase in reinforcement strength leads to fuller hysteresis loops, indicating enhanced yield and peak loads.

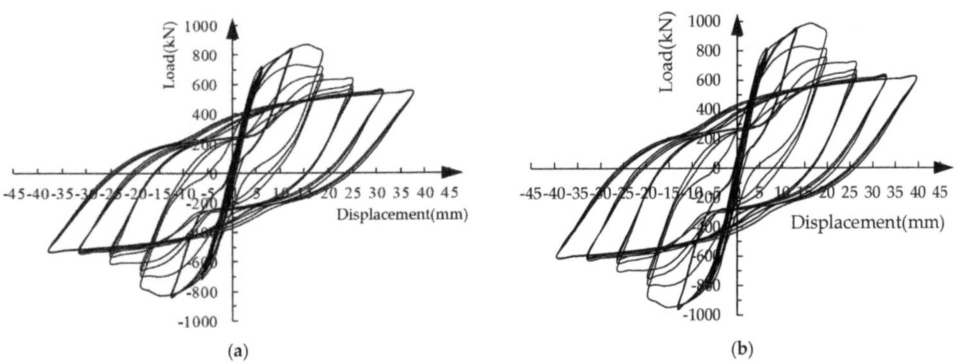

Figure 19. *Cont.*

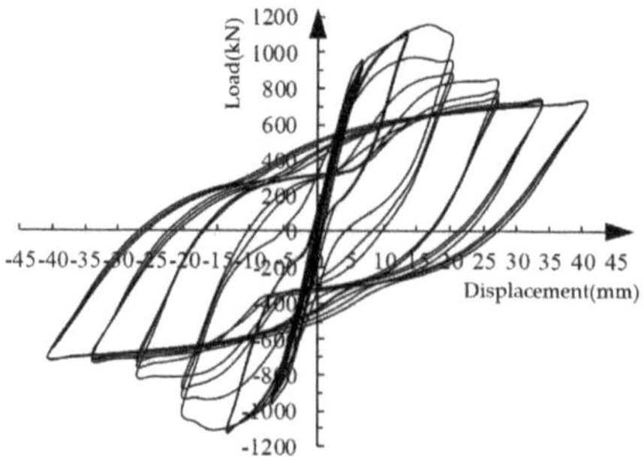

(c)

Figure 19. Load–displacement hysteresis curves with varied reinforcement strengths. (**a**) Structure 5-1; (**b**) structure 5-2; (**c**) structure 5-3.

4.5.3. Energy Dissipation Capacity

Table 14 displays the energy dissipation values for structural specimens with different strengths of longitudinal reinforcement. The data reveals a clear trend: structure 5-3 > structure 5-2 > structure 5-1, both in terms of single-cycle and cumulative energy dissipation. This indicates an increase in energy dissipation capacity with the enhancement of longitudinal reinforcement strength. However, high-strength steel bars contribute more to the structural performance before yield. Considering the high price and the reduction in ductility, using high-strength steel is not essential, and the HRB400 steel bar is a better choice.

Table 14. Energy dissipation with varied reinforcement strengths (J).

Displacement Loading Level	Number of Cycles	Structure 5-1		Structure 5-2		Structure 5-3	
		Single-Cycle	Cumulative	Single-Cycle	Cumulative	Single-Cycle	Cumulative
1Δ	1	1551.2	1551.2	1797.9	1797.9	2079.2	2079.2
	2	738.9	2290.1	851.8	2649.7	980.2	3059.4
	3	435.2	2725.3	496.2	3145.9	569.3	3628.7
2Δ	1	7502.7	10,228	8679.7	11,825.6	10,056.5	13,685.2
	2	6475.0	16,703	7490.8	19,316.4	8682.3	22,367.5
	3	6162.1	22,865.1	7128.8	26,445.2	8269.4	30,636.9
3Δ	1	19,152.0	42,017.1	22,147.1	48,592.3	25,690.6	56,327.5
	2	18,233.0	60,250.1	21,093.2	69,685.5	24,468.1	80,795.6
	3	15,311.9	75,562	17,704.6	87,390.1	20,531.3	101,326.9
4Δ	1	24,264.6	99,826.6	28,071	115,461.1	32,562.4	133,889.3
	2	22,072.1	121,898.7	25,534.6	140,995.7	29,612.1	163,501.4
	3	21,431.1	143,329.8	24,787.2	165,782.9	28,753.2	192,254.6
5Δ	1	31,878.6	175,208.4	36,879.5	202,662.4	42,780.2	235,034.8
	2	31,681.2	206,889.6	36,651.1	239,313.5	42,510.3	277,545.1
	3	31,019.4	237,909	35,885.5	275,199	41,627.2	319,172.3
6Δ	1	41,156.9	279,065.9	47,613.3	322,812.3	55,227.4	374,399.7
	2	39,819.8	318,885.7	46,066.4	368,878.7	53,429.0	427,828.7
	3	39,376.4	358,262.1	45,537.2	414,415.9	52,819.2	480,647.9

5. Results and Optimization Design Suggestions

The ECC/RC composite shear wall structure has a good seismic energy dissipation capacity, and ECC is as easy to install as concrete; however, it also has some limitations. For example, it is difficult to repair when the shear walls are damaged. In addition, the price of ECC is about 2.5 times that of concrete, and the high cost makes ECC unable to be widely used in the ECC/RC composite shear wall structures. Therefore, it is necessary to optimize the design of the prefabricated ECC/RC composite shear wall structure, considering its technical performance and cost.

Machine learning, artificial intelligence, and neural network methods [45–48] have significant advantages in optimal design. Purohit et al. [49] used deep learning technology to obtain effective segmentation results. Zhao et al. [50] proposed an intelligent design method for a beam and slab of shear wall structure based on deep learning. Du et al. [51] established a rapid optimization method for flexible support structures based on mathematical models.

This paper draws on machine learning, artificial intelligence, and neural network methods. Emphasis is placed on critical aspects, such as the optimal deployment areas for ECC within composite coupling beams and shear walls, the grade of ECC strength, the proportion of stirrups in coupling beams, and the caliber of longitudinal reinforcement. Through finite element analysis, this research quantitatively assesses the impact of these variables on seismic energy dissipation, incorporating evaluations of load–displacement hysteretic behaviors and the energy dissipation potential of ECC/RC shear wall samples. Insights from this analysis reveal the following:

- Incorporation of ECC in composite coupling beams significantly bolsters seismic resilience compared to traditional concrete counterparts, with benefits amplifying alongside increased ECC integration.
- The most effective energy dissipation is achieved with ECC applied 400 mm up from the shear wall's base, recommending a 14% structural height allocation (400 mm) for optimal ECC integration in prefabricated walls.
- ECC materials of lesser strength demonstrate enhanced energy dissipation abilities under prolonged loading conditions.
- Variations within the examined stirrup ratio spectrum (0.5% to 2.01%) have a minimal effect on the seismic performance.
- The necessity for high-strength steel is de-emphasized, with HRB400 grade steel emerging as the preferable option.

Based on the above research results, considering both cost-efficiency and performance, the study advocates for strategic ECC deployment within coupling beams and recommends a 14% structural elevation (400 mm) at shear walls' base. Optimal parameters proposed include ECC strength grade E40, a longitudinal reinforcement of HRB400, and a stirrup ratio within coupling beams set at 0.5%, detailed in Table 15.

Table 15. Recommended parameter values.

Use Regions of ECC in Composite Coupling Beams	Use regions of ECC in Shear Walls	Stirrup Ratio of Coupling Beams	Strength of Longitudinal Reinforcement	Strength of ECC
Coupling beams	14% structural height (400 mm)	0.5%	HRB400	E40

6. Discussion

In this paper, the seismic energy dissipation performance of the prefabricated ECC/RC composite shear wall structure is studied by means of experiments and numerical simulation. Based on the experimental verification of numerical simulation, this study employs finite element analysis to examine the impact of various factors on seismic energy dissipation under low cyclic loading. The study focuses on key areas, such as the application zones of ECC in composite coupling beams and shear walls, ECC strength, the stirrup ratio in coupling beams, and the strength of longitudinal reinforcement. The influence of these parameters on seismic energy dissipation is quantitatively evaluated by finite element analysis. These analyses include load–displacement hysteretic curves, as well as an assessment of the energy dissipation capacity of ECC/RC shear wall specimens. Based on the above research results, the optimization design recommendations are put forward.

Reflecting on the analysis of a half-scale, two-story spatial structure that closely replicates an actual test specimen, this research offers insightful guidelines for engineering applications. Nonetheless, it emphasizes the need for further exploration into the behavior of such engineering solutions under real earthquake conditions to ensure wide-ranging applicability. In the future, we will analyze the seismic performance of full-scale engineering structures, and use the methods of artificial intelligence, machine learning, and neural networks to conduct more accurate optimization design research on practical engineering applications.

Author Contributions: Conceptualization, J.Y. and H.G.; Methodology, J.Y. and R.Z.; Software, J.Y. and R.Z.; Validation, M.S.; Formal analysis, J.Y., M.S. and G.Y.; Investigation, J.Y. and H.G.; Resources, G.Y.; Data curation, J.Y. and R.Z.; Writing—original draft, J.Y.; Writing—review & editing, M.S.; Visualization, R.Z.; Supervision, M.S. and R.Z.; Project administration, G.Y. and H.G.; Funding acquisition, J.Y., M.S., G.Y. and H.G. All authors have read and agreed to the published version of the manuscript.

Funding: The research was funded by the "National Natural Science Foundation of China" (Grant No. 52308271), "Shenzhen Natural Science Foundation"(Grant No. JCYJ20230807114401002), "Shenzhen High-level Talents Research Start-up Fund (Grants No. RC2022-004)"and the "2022 Doctoral Research Startup Project of Shenzhen Institute of Information Technology" (Grant No. SZIIT2022KJ014), supported by "The Eighth Batch of Education and Teaching Reform Research Projects of Shenzhen Institute of Information Technology" (Grant No. 2021dbpjgyb25), and by "Guangdong Province Ordinary University Features Innovative Projects" (Grants No. 2020 ktscx300 and 2023KTSCX321).

Data Availability Statement: The original contributions presented in the study are included in the article, further inquiries can be directed to the corresponding authors.

Conflicts of Interest: The authors declare no conflict of interest.

References

1. Elliott, K.S. *Precast Concrete Structures*; CRC Press: Boca Raton, FL, USA, 2016.
2. Soares, D.; De Brito, J.; Ferreira, J.; Pacheco, J. Use of coarsere cycled aggregates from precast concrete rejects: Mechanical and durability performance. *Construct. Build. Mater.* **2014**, *71*, 263–272. [CrossRef]
3. Kurama, Y.C.; Sritharan, S.; Fleischman, R.B.; Restrepo, J.I.; Henry, R.S.; Cleland, N.M.; Ghosh, S.K.; Bonelli, P. Seismic-resistant precast concrete structures: State of the art. *J. Struct. Eng.* **2018**, *144*, 03118001. [CrossRef]
4. Shen, S.D.; Cui, Y.; Pan, P.; Ren, J.Y. Development of prefabricated composite energy dissipating slotted shear wall. *Eng. Struct.* **2019**, *199*, 109577. [CrossRef]
5. Nazari, M.; Sritharan, S.; Aaleti, S. Single precast concrete rocking walls as earthquake force-resisting elements. *Earthq. Eng. Struct. Dynam.* **2016**, *46*, 753–769. [CrossRef]
6. Peng, Y.Y.; Qian, J.R.; Wang, Y.H. Cyclic performance of precast concrete shear walls with a mortar–sleeve connection for longitudinal steel bars. *Mater. Struct.* **2016**, *49*, 2455–2469. [CrossRef]
7. Brunesi, E.; Peloso, S.; Pinho, R.; Nascimbene, R. Cyclic testing of a full-scale two-storey reinforced precast concrete wall-slab-wall structure. *Bull. Earthq. Eng.* **2018**, *16*, 5309–5339. [CrossRef]
8. Brunesi, E.; Peloso, S.; Pinho, R.; Nascimbene, R. Shake-table testing of a full-scale twostorey precast wall-slab-wall structure. *Earthq. Spectra* **2019**, *35*, 1583–1609. [CrossRef]
9. Qian, K.; Li, B.; Liu, Y. Integrity of precast concre test ructues to resist progressive collapse. *Geotech. Struct. Eng. Congr.* **2016**, 1976–1986. [CrossRef]
10. Wu, B.; Peng, C.W.; Zhao, X.Y. Cyclic loading tests of semi-precast circular steel tubular columns in corporating precast segments containing demolished concrete lumps. *Eng. Struct.* **2020**, *211*, 110438. [CrossRef]
11. Qian, J.R.; Han, W.L.; Zhao, Z.Z.; Qin, H.; Zhang, Y.; Yu, J.; Ma, T.; Tian, D. Pseudo-dynamic substructure test on a 3-storey full-scale model of prefabricated concrete shear wall structure with rebars splicing by grout sleeves. *J. Build. Struct.* **2017**, *38*, 26–38. (In Chinese)
12. *SJG98-2021;* Technical Specification for Concrete Structures of Tall Building. China Architecture & Building Press: Beijing, China, 2021.
13. Xiao, S.; Wang, Z.; Li, X.; Harries, K.A.; Xu, Q.; Gao, R. Study of effects of sleeve grouting defects on the seismic performance of precast concrete shear walls. *Eng. Struct.* **2021**, *236*, 111833. [CrossRef]
14. Zhi, Q.; Kang, L.; Jia, L.; Xiong, J.; Guo, Z. Seismic performance of precast shear walls prestressed via post-tensioned high strength bars placed inside grouted corrugated pipes. *Eng. Struct.* **2021**, *237*, 112153. [CrossRef]
15. Tan, W.; Xianqian, M.; Ruinian, J.; Lijun, D.; Jun, L.; Yinlong, H.; Haibin, M. Seismic performance of repaired prefabricated shear walls. *Case. Stud. Constr. Mat.* **2023**, *18*, e01896. [CrossRef]
16. Yang, J.; Guo, T.; Wang, W.; Zhu, R. Experimental investigation on seismic behaviors of precast concrete shear walls with sleeve grouting defects. *Adv. Struct. Eng.* **2023**, *26*, 258–271. [CrossRef]

17. Qin, C.; Bai, G.; Liu, B.; Wu, T.; Wang, B. Study on the global bidirectional seismic behavior of monolithic prefabricate concrete shear wall structure. *Soil Dyn. Earthq. Eng.* **2020**, *136*, 106194. [CrossRef]
18. Sheng, F. *Study on Damping Characteristics and Dynamic Response of ECC/RC Prefabricated Shear Wall Structure*; Southeast University: Nanjing, China, 2022. (In Chinese)
19. Li, V.C. Engineered cementitious composites (ECC) material, structural, and durability performance. In *Concrete Construction Engineering Handbook*; Nawy, E., Ed.; CRC Press: Boca Raton, FL, USA, 2008; Chapter 24; pp. 1–78.
20. Yu, J.; Yao, J.; Lin, X.; Li, H.; Lam, J.Y.; Leung, C.K.; Shih, K. Tensile performance of sustainable Strain-Hardening Cementitious Composites with hybrid PVA and recycled PET fibers. *Cem. Concr. Res.* **2018**, *107*, 110–123. [CrossRef]
21. Yu, K.Q.; Yu, J.T.; Dai, J.G.; Lu, Z.D.; Shah, S.P. Development of ultra-high performance engineered cementitious composites using polyethylene (PE) fibers. *Constr. Build. Mater.* **2018**, *158*, 217–227. [CrossRef]
22. Abd Elmoaty, A.E.M.; Morsy, A.M.; Harraz, A.B. Effect of Fiber Type and Volume Fraction on Fiber Reinforced Concrete and Engineered Cementitious Composite Mechanical Properties. *Buildings* **2022**, *12*, 2108. [CrossRef]
23. Abouhussien, A.A.; Hassan, A.A.; Ismail, M.K.; AbdelAleem, B.H. Evaluating the cracking behavior of ECC beam-column connections under cyclic loading by acoustic emission analysis. *Constr. Build. Mater.* **2019**, *215*, 958–968. [CrossRef]
24. Hao, Z.; Lu, C.; Li, Z. Highly accurate and automatic semantic segmentation of multiple cracks in engineered cementitious composites (ECC) under dual pre-modification deep-learning strategy. *Cem. Concr. Res.* **2023**, *165*, 107066. [CrossRef]
25. Huang, B.T.; Wu, J.Q.; Yu, J.; Dai, J.G.; Leung, C.K.; Li, V.C. Seawater sea-sand engineered/strain-hardening cementitious composites (ECC/SHCC): Assessment and modeling of crack characteristics. *Cem. Concr. Res.* **2021**, *140*, 106292. [CrossRef]
26. Gu, D.; Pan, J.; Mustafa, S.; Huang, Y.; Luković, M. Shear transfer mechanism in reinforced engineered cementitious composite (ECC) beams: Quantification of Vs and Vc. *Eng. Struct.* **2022**, *261*, 114282. [CrossRef]
27. Meng, D.; Lee, C.K.; Zhang, Y.X. Structural behavior of reinforced polyvinyl alcohol engineered cementitious composite (PVA-ECC) beams under static and fatigue loadings. In *Advances in Engineered Cementitious Composites*; Woodhead Publishing: Sawston, UK, 2022; pp. 161–208.
28. Abbas, A.A.; Arna'Ot, F.H.; Abid, S.R.; Özakça, M. Flexural behavior of ECC hollow beams incorporating different synthetic fibers. *Front. Struct. Civ. Eng.* **2021**, *15*, 399–411. [CrossRef]
29. Amiri, M.; Esfahani, M.R. Effect of using Engineered Cementitious Composites (ECC) on failure behavior of flat slab-column connections. *Structures* **2023**, *47*, 2397–2407. [CrossRef]
30. Zhang, Z.; Qian, S.; Ma, H. Investigating mechanical properties and self-healing behavior of micro-cracked ECC with different volume of fly ash. *Constr. Build. Mater.* **2014**, *52*, 17–23. [CrossRef]
31. Zhang, Z.Y.; Ding, R.; Fan, J.S.; Nie, X.; Zhang, J. Seismic performance and shear strength of coupling beams using engineered cementitious composites with different reinforcement layouts. *Eng. Struct.* **2020**, *219*, 110895. [CrossRef]
32. Suryanto, B.; Tambusay, A.; Suprobo, P.; Bregoli, G.; Aitken, M.W. Seismic performance of exterior beam-column joints constructed with engineered cementitious composite: Comparison with ordinary and steel fibre reinforced concrete. *Eng. Struct.* **2022**, *250*, 113377. [CrossRef]
33. Cai, J.; Pan, J.; Xu, L.; Li, G.; Ma, T. Mechanical behavior of RC and ECC/RC composite frames under reversed cyclic loading. *J. Build. Eng.* **2021**, *35*, 102036. [CrossRef]
34. Khan, F.A.; Khan, S.W.; Khan, W. Seismic performance of RC-ECC composite frame by eliminating shear reinforcement in beam-column joint: Shake table tests. *Adv. Struct. Eng.* **2022**, *25*, 2966–2980. [CrossRef]
35. Khan, F.A.; Rashid, M.; Khan, S.W.; Rizwan, M.; Badrashi, Y.I.; Fahim, M.; Gul, A. Comparative seismic performance assessment of RC and RC/ECC hybrid frame structures: A shake table study. *Innov. Infrastruct. Solut.* **2022**, *7*, 94. [CrossRef]
36. Ye, B.; Wang, H.; Ma, Y.; Pan, P. Seismic performance of flexure-dominated reinforced-engineered cementitious composites coupled shear wall. *Eng. Struct.* **2022**, *272*, 114992. [CrossRef]
37. Yang, J.; Liang, S.; Zhu, X.; Sun, C.; Guo, Z. Seismic behavior of precast concrete coupled shear walls with engineered cementitious composite (ECC) in the critical cast-in-place regions. *Sci. China Technol. Sci.* **2017**, *60*, 1244–1254. [CrossRef]
38. Yang, J.; Jiang, L.; Guo, H.; Yao, G. Stress Mechanism and Energy Dissipation Performance Optimization of Prefabricated ECC/RC Combined Shear Walls under Low Cyclic Loading. *Buildings* **2023**, *13*, 772. [CrossRef]
39. *JGJ 1-2014*; Technical Specification for Precast Concrete Structures. China Architecture & Building Press: Beijing, China, 2014.
40. Hou, H.; Qiu, C.; Li, G.; Wang, J. Cyclic test on steel frames with energy-saving sandwich composite panels. *Eng. Mech.* **2012**, *9*, 177–184. (In Chinese)
41. *GB 50010-2010*; Code for Design of Concrete Structures (2015 Edition). China Architecture & Building Press: Beijing, China, 2015.
42. Yuan, F.; Pan, J.; Wu, Y. Numerical study on flexural behaviors of steel reinforced engineered cementitious composite (ECC) and ECC/concrete composite beams. *Sci. China Technol. Sci.* **2014**, *57*, 637–645. [CrossRef]
43. Zhang, J.; Wang, Q.; Hu, S.; Wang, C. Parameters verification of concrete damaged plastic model of ABAQUS. *Build. Struct.* **2008**, *38*, 127–130.
44. Kien, D.N.; Chen, X.; Zhuang, X.; Rabczuk, T. Radial Basis Function Based Finite Element Method for Bending, Vibration and Buckling Analysis of Laminated Composite Mindlin-Reissner Plates. In *Advances in Engineering Research and Application, Proceedings of the International Conference on Engineering Research and Applications, ICERA 2022, Thai Nguyen, Vietnam, 1–2 December 2022*; Springer International Publishing: Cham, Switzerland, 2022; pp. 806–822.

45. Gheisari, M.; Ebrahimzadeh, F.; Rahimi, M.; Moazzamigodarzi, M.; Liu, Y.; Dutta Pramanik, P.K.; Ali Heravi, M.; Mehbodniya, A.; Ghaderzadeh, M.; Feylizadeh, M.R.; et al. Deep learning: Applications, architectures, models, tools, and frameworks: A comprehensive survey. *CAAI Trans. Intell. Technol.* **2023**, *8*, 581–606. [CrossRef]
46. Koy, A.; Çolak, A.B. The Intraday High-Frequency Trading with Different Data Ranges: A Comparative Study with Artificial Neural Network and Vector Autoregressive Models. *Arch. Adv. Eng. Sci.* **2023**, 1–20. [CrossRef]
47. Li, Z.; Li, S. Recursive recurrent neural network: A novel model for manipulator control with different levels of physical constraints. *CAAI Trans. Intell. Technol.* **2023**, *8*, 622–634. [CrossRef]
48. Zhao, L.; Sun, Z.; Liu, K.; Zhang, J. The dynamic relaxation form finding method aided with advanced recurrent neural network. *CAAI Trans. Intell. Technol.* **2023**, *8*, 635–644. [CrossRef]
49. Purohit, J.; Dave, R. Leveraging deep learning techniques to obtain efficacious segmentation results. *Arch. Adv. Eng. Sci.* **2023**, *1*, 11–26. [CrossRef]
50. Zhao, P.; Liao, W.; Xue, H.; Lu, X. Intelligent design method for beam and slab of shear wall structure based on deep learning. *J. Build. Eng.* **2022**, *57*, 104838. [CrossRef]
51. Du, Q.; Luo, G.; Wang, X.; Wang, T.; Fu, G.; Lu, C. Fast Optimization Method of Flexible Support Structure Based on Mathematical Model. *Arch. Adv. Eng. Sci.* **2023**, 1–11. [CrossRef]

buildings

Article

Experimental Studies and Finite Element Analysis of Socket-Type Keyway Steel Pipe Scaffolding

Chenyang Zhang [1,2], Jianjun Yang [1,2], Liqiang Jiang [1,2,*] and Yanqing He [3]

1 School of Civil Engineering, Central South University, Changsha 410075, China; chenyoung9844@163.com (C.Z.); jjyang@csu.edu.cn (J.Y.)
2 National Engineering Laboratory for High-Speed Railway Construction, Changsha 410075, China
3 Baoji Works Section, China Railway Xi'an Bureau Group Co., Ltd., Baoji 721000, China; hyq891219@163.com
* Correspondence: jianglq2019@csu.edu.cn

Abstract: Scaffolding is an integral temporary structural system in the field of construction engineering. However, the current scaffolding commonly has the shortcomings of low construction efficiency and high risk. This paper proposes a novel socket-type keyway steel pipe scaffolding, which can well solve the shortcomings of the existing scaffolding. Due to less research related to scaffolding in the past decades, it has resulted in a high number of scaffolding accidents. In order to avoid the occurrence of scaffolding accidents, it is necessary to systematize the study of this novel type of scaffolding. This study is an extremely important reference for the use and design of this novel type of scaffolding. To explore the ultimate load capacity and destabilization mode of the novel socket-type keyway steel pipe scaffolding, full-scale tests were conducted on the socket-type keyway steel pipe scaffolding with cantilever heights of 1.2 m and 0.5 m. The test results indicate that the ultimate load capacity of the scaffolding with a cantilever height of 1.2 m is 196 kN, and the destabilization mode is local instability. The ultimate load capacity with a cantilever height of 0.6 m is 276 kN, and the destabilization mode is half-wave buckling. This phenomenon shows that the different cantilever heights of the scaffolding have a significant effect on the load capacity and destabilization mode. Moreover, the load capacity decreases significantly with increasing cantilever length. The finite element model was established using SAP2000 v21 and compared with the test results. The error between the ultimate load capacity in the finite element linear elastic buckling analysis and the test results is 25%. The error between the calculated ultimate load capacity in the nonlinear buckling analysis considering the initial geometrical defects and the test results is 4%. Therefore, the nonlinear buckling analysis considering the initial geometrical defects is more in line with the force situation of the structure in the real situation.

Keywords: socket-type keyway; ultimate load capacity; finite element analysis; buckling analysis; failure mechanism

Citation: Zhang, C.; Yang, J.; Jiang, L.; He, Y. Experimental Studies and Finite Element Analysis of Socket-Type Keyway Steel Pipe Scaffolding. *Buildings* **2024**, *14*, 245. https://doi.org/10.3390/buildings14010245

Academic Editor: Oldrich Sucharda

Received: 9 December 2023
Revised: 8 January 2024
Accepted: 13 January 2024
Published: 16 January 2024

1. Introduction

Beginning in the mid-1980s, infrastructure development has been taking place throughout China, necessitating the use of modern construction techniques. Traditional scaffolding has proven inadequate in meeting the demands of this rapid development. As a result, many types of steel pipe scaffolding have begun to appear [1]. These include fastener-type steel pipe scaffolding [2], portal-type steel pipe scaffolding [3], bowl buckle-type steel pipe scaffolding [4], disc-type steel pipe scaffolding [5], and many other types of scaffolding [6]. After the 1990s, the building structure gradually developed in the direction of a large span, high tower, and heavy load. The existing scaffolding cannot be adapted to the needs of building construction. Therefore, based on the traditional scaffolding, improved novel scaffolding gradually emerged. The socket-type keyway steel pipe scaffolding is one of them. As a novel type of scaffolding, it has been gradually used in various projects due to its advantages of safe and reliable construction, fast dismantling speed, and good economic returns.

The uniqueness of the novel type of socket-type keyway steel pipe scaffolding is reflected in the upright rod welded with a keyway socket that can be connected in four directions. The horizontal rod ends are machined directly into keyway plugs to be vertically inserted into the sockets of the upright rod. The upright rods are connected by a trocar. Horizontal rods and diagonal rods are quickly connected using rod-end keyed plugs that snap into keyway sockets to form a stable structural geometrically invariant system, as shown in Figure 1. The nodes of the rack are self-locked by the friction of the contact surfaces between the plug and the socket, and the reliability of the connection is ensured. Compared with other steel pipe scaffoldings, this kind of steel pipe scaffolding has a simple erection process. The upright rods and horizontal rods are connected by sockets and plugs, which makes the installation and dismantling of the scaffolding quick and easy. The center axes of the horizontal rods and upright rods coincide so that the node force transmission is not affected by the eccentric force. In addition, the socket of the bracket is directly welded to the upright rod, which provides good stability, high load-bearing capacity, and fittings that are not easily lost.

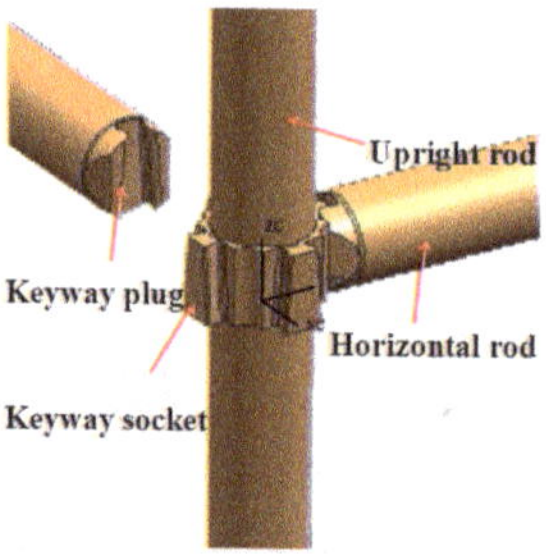

Figure 1. The structural layout of the all-steel attached lifting protection platform.

The scaffolding market and technology are growing by leaps and bounds. The lack of corresponding research and the imperfections of relevant specifications have led to a series of engineering accidents [7–10]. Scholars have gradually emphasized the systematic study of various types of scaffolding. Regarding fastener-type steel pipe scaffolding. Jia [11] analyzed the destabilization mechanism and damage mode of an ultra-high full room of a fastener steel pipe scaffold under uniform loads through static tests of seven models and analyzed the effect of multiple parameters on the ultimate load of a full scaffold. According to the finite element analysis, the suggested value of the coefficient of unevenness for the calculation of the slip-resistant bearing capacity of the fasteners at the top level of the full scaffold is 1.51. In Ji [12], according to the joint stiffness value determined by the experimental study and the theory of semirigid connection frame with lateral displacement, the calculation formulas of the stiffness correction coefficient of the transverse rod and the constraint coefficient at the end of the vertical rod are derived, and the effective length coefficient of the vertical rod and the theoretical value of the stable bearing capacity of the vertical rod under different working conditions are given. Liu [13] compared the tests with the ANSYS model by connecting fastener-type scaffolds with different nodes. He analyzed the strength and damage forms of wheel-buckle scaffolds with different nodes and determined the most favorable node connections. Dong [14] analyzed the structural stability of a wheel-buckled steel pipe scaffolding. The results indicated that the damage of the single-layer upright rods is typical with lateral displacement buckling, while the damage of the double-layer upright rods is without lateral displacement buckling. Yu [15] investigated the failure mode and stable bearing capacity of a novel type of wheel-buckle scaffolding through nine full-size test models and established a simplified formula for the stabilizing load capacity of the scaffolding. For the novel type of scaffolding, Chen [16] developed a kind of superstrong thin-walled steel pipe fastener scaffold. When the weight of the superstrong thin-walled steel pipe is reduced by 40%, there is no significant decrease in the vertical load capacity of the steel pipe joist. Bian [17] performed mechanical test research on a new type of wheel buckle-type steel pipe scaffolding. The ultimate load capacities of pins in eccentric tension and horizontal rods in bending, as well as the damaged forms of wheel discs in compression, sleeves in shear, and upright rods in instability, were derived from the tests. In addition, scholars have been more concerned with the connecting nodes of scaffoldings. Pienko [18] determined the load capacity of a disc-buckled scaffold by applying loads in different directions of action to the nodes. Zhang [19] used a finite element model to analyze the rotational stiffness of the connection nodes of disc-buckled scaffolds under different connection forms and loading modes. Thus, the load-bearing mechanism of the nodes of the disc-buckled scaffolds was summarized.

As a novel type of scaffolding system, the socket-type keyway steel pipe scaffolding has been applied in some projects. However, there are fewer theoretical studies on this novel type of scaffolding. Only a few scholars have studied and analyzed the node force performance [20,21] and rotational stiffness [22]. Therefore, this paper carries out a full-frame foot test on two types of socketed keyway steel pipe scaffoldings with different cantilever heights, which have been used in actual projects. The finite element software SAP2000 v21 is used to establish the test model and compare it with the test results to study the ultimate load capacity and damage mode of the novel type of scaffolding.

2. Overview of the Experiment

2.1. Experimental Design

A full-scale test was carried out on two types of socket-type keyway steel pipe scaffoldings with different cantilever heights, which were used to study the damage pattern and ultimate load capacity of this novel type of scaffolding. The overall arrangement of the two types of steel pipe scaffoldings is shown in Figure 2. The specifications of the component materials used in the tests are shown in Table 1. The upright rods are connected by a trocar. Horizontal rods and diagonal rods are connected using rod-end keyed plugs. The detail of the rod-end connection is shown in Figure 3. In the two test schemes, the spacing of the upright rod is 1 m, and the step distance of the upright rod is 1.5 m or 1 m. The elongation length of the top cantilever end of scheme I is 1.2 m, while the elongation length of the top cantilever end of scheme II is 0.65 m. The test schemes of the whole scaffolding are shown in Table 2.

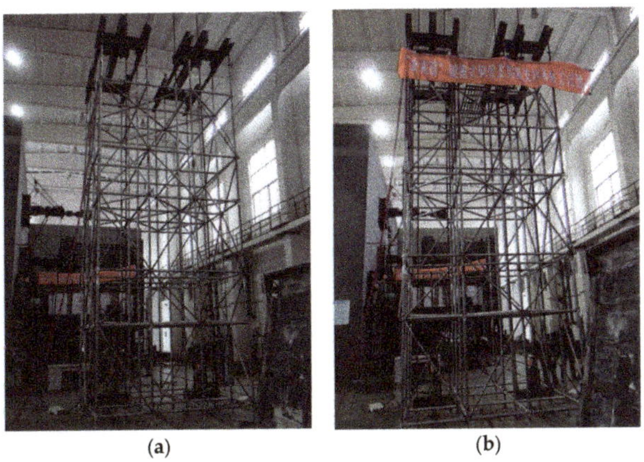

Figure 2. The overall layout of the test model. (**a**) Scheme I. (**b**) Scheme II.

Table 1. Specification of component materials.

Component	Section Form	Norm (mm)	Thicknesses (mm)	Material
Upright rod	Hot-dip galvanized welded steel pipe	Ø 48	3.0	Q235
Horizontal rod	Welded steel pipe	Ø 48	3.0	Q235
Diagonal rod	Round steel tube	Ø 32	2.0	Q235

(**a**) (**b**)

Figure 3. Detail of rod-end connection. (**a**) Connection of upright rods with horizontal rods and diagonal rods. (**b**) Connection of upright rods.

Table 2. Model parameters of the scaffolding test.

Experiment Scheme	Specifications	Horizontal Diagonal Brace	Vertical Brace	Length of Cantilever End
I	Three hurdles and six steps	Three-layer setup	Surrounding facades	1.2 m
II	Three hurdles and six steps	Three-layer setup	Surrounding facades	0.65 m

2.2. Loading Program and Measurement Point Arrangement

The test was carried out by self-balancing loading. Uniform and symmetrical loading of the whole scaffolding was applied by two 100 t jacks at the bottom, reaction beams, and a double-layer distribution beam at the top. The loading device for the test is shown in Figure 4. The applied load is 10 kN per stage for scheme I and 20 kN for scheme II. Each level of loading lasted for 10 min. Eventually, the entire scaffolding was destroyed.

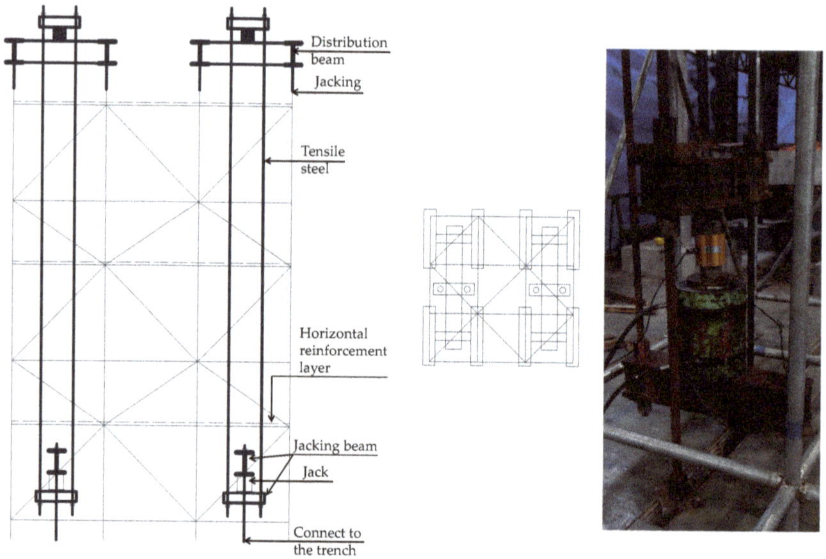

(**a**) (**b**) (**c**)

Figure 4. Loading setup diagram. (**a**) Side view. (**b**) Top view. (**c**) Detail of loading device.

The arrangement of the strain measurement points is shown in Figure 5. Measurement points 1, 3, 5, and 7 are located on the south side of the cantilever end of the upright rod. Measurement points 2, 4, 6, and 8 are located on the east side of the cantilever end of the upright rod. Measurement points 9, 10, and 11 are located in the middle of the uppermost horizontal rod. Measurement points 12 and 13 are located in the middle of the uppermost diagonal rod.

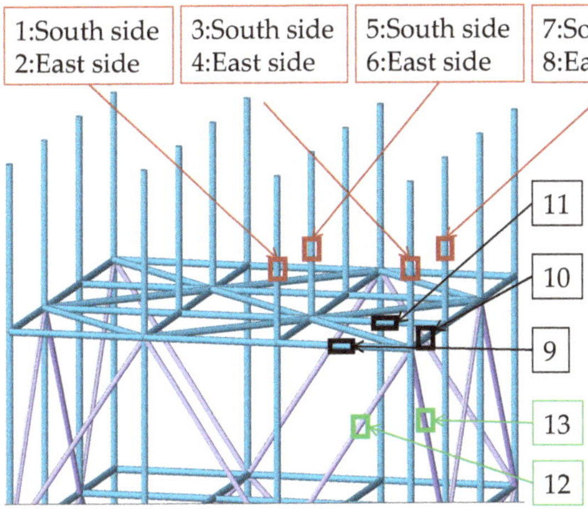

Figure 5. Arrangement of the strain measuring points in schemes I and II.

3. Experimental Results and Analysis

3.1. Experimental Phenomena

Scheme I: As shown in Figure 6a, due to the failure of the loading equipment, the two rows of upright rods on the north side of the scaffolding broke down when the load was 130 kN. The two rows of upright rods on the south side of cantilever end buckled when loaded to 160 kN (Figure 6b). At 180 kN, the bending deformation at the lower end node of the cantilever end rapidly developed, and the cantilever end tilted northward. The ultimate load capacity of the final frame damage was 196 kN, and the instability mode was localized instability (Figure 7a).

(**a**) (**b**)

Figure 6. Instability morphology of scheme I. (**a**) West elevation. (**b**) East elevation.

Scheme II: As shown in Figure 8a, there was no significant deformation of the frame when the load was 140 kN. The middle four upright rods exhibited slight deformation when loaded to 200 kN (Figure 8b). Eventually, the cantilever end bends to the south when loaded to 260 kN. The middle four upright rods bent to the north at the fourth to sixth (between the horizontal reinforcement layers) levels of the horizontal rods. The whole frame was destabilized by half-wave drumming (Figure 7b).

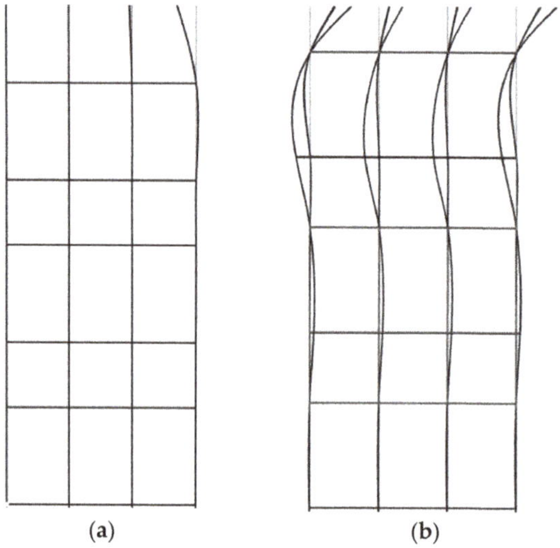

Figure 7. Instability mode of schemes I and II. (**a**) Instability mode of scheme I;. (**b**) Instability mode of scheme II.

Figure 8. Instability morphology of scheme II. (**a**) West elevation. (**b**) South elevation.

3.2. Load-Strain Curve

The load-strain curves at each measurement point of scheme I and scheme II in the test are shown in Figures 9 and 10, respectively. The strains of the upright rods were close to 0.1ε in both schemes when the steel pipe scaffolding suffered destabilizing damage. The lower end of the cantilever end of the upright rod reached the yield stage when the scaffolding was damaged, while the strains in the horizontal rod and the vertical diagonal brace were low. This is because the scaffolding was mainly subjected to vertical loads, and the rods were mainly subjected to axial forces. Therefore, under the action of a vertical load, vertical rods were subjected to larger loads, while horizontal rods and vertical diagonal braces were subjected to smaller forces. Thus, the strains measured in the horizontal rods and vertical diagonal braces were small compared to the strains in the upright rods.

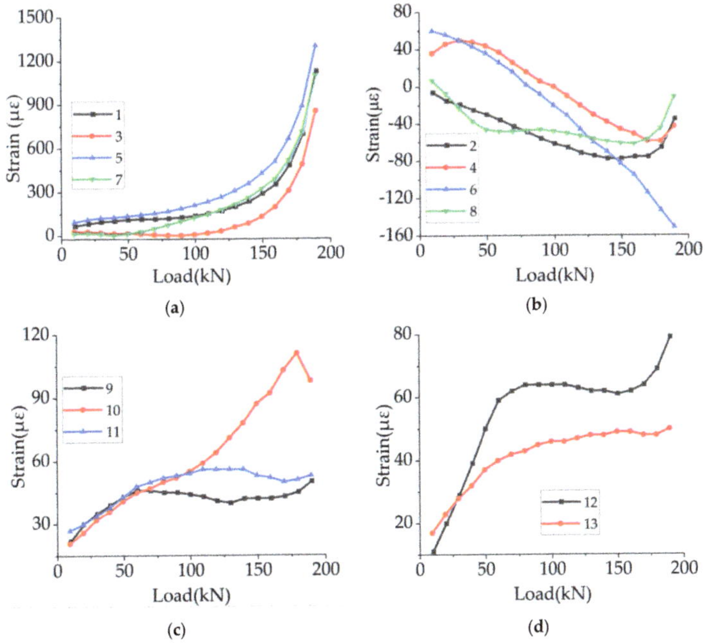

Figure 9. Load—strain curves at the measurement point points in scheme I. (**a**) Upright rod load-strain diagram 1. (**b**) Upright rod load-strain diagram 2. (**c**) Horizontal rod load-strain diagram. (**d**) Diagonal rod load-strain diagram.

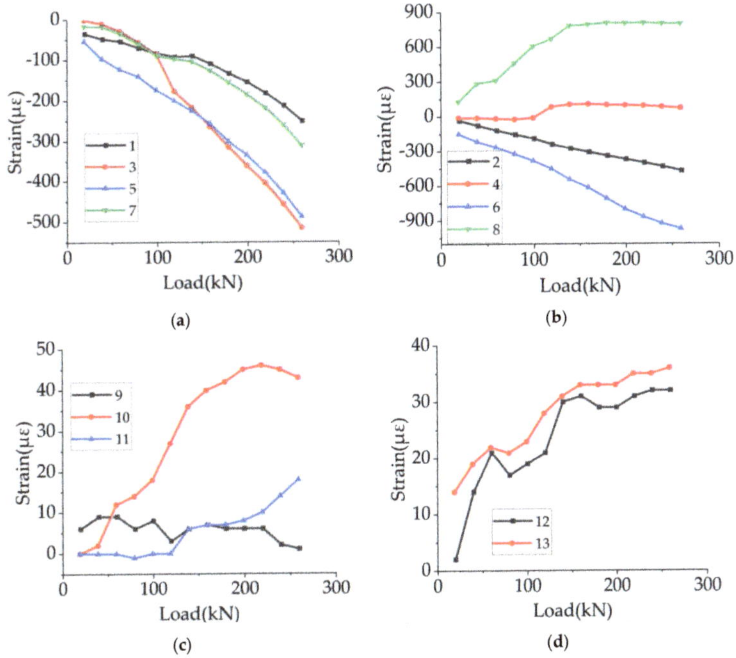

Figure 10. Load—strain curves at the measurement points in scheme II. (**a**) Upright rod load-strain diagram 1. (**b**) Upright rod load-strain diagram 2. (**c**) Horizontal rod load-strain diagram. (**d**) Diagonal rod load-strain diagram.

The two rows of upright rods on the north side of the scaffolding in scheme I broke down when the load was 130 kN (Figure 6a). Therefore, the test results do not take into account the two rows of upright rods on the north side. The final form of damage was a localized instability damage of the two rows of upright rods on the south side of the top tilted to the north side (Figures 6b and 8a). The strains at measurement points 1, 3, 5, and 7 are much greater than the strains at measurement points 2, 4, 6, and 8 (Figure 9a,b).

The final form of damage in scheme II is overall destabilization damage. The cantilever ends of the upright rods deform in both directions (Figure 7a,b). Therefore, the strains at measurement points 1, 3, 5, and 7 of scheme II are closer to the values of the strains at measurement points 2, 4, 6, and 8 of scheme II (Figure 10a,b), and the strains in scheme II are smaller than the strains at measurement points 1, 3, 5, and 7 of scheme I (Figure 9a).

3.3. *Analysis of Test Results*

The length of the cantilever end of the scaffolding in scheme I is longer. The localized instability at the cantilever end resulted in the entire scaffolding not being able to continue loading. The length of the cantilever end of the scaffolding in scheme II is half the length of the cantilever end of the frame in scheme I. The ultimate load capacity of scheme II is 33% greater than that of scheme I. The final damage form of the scaffolding of scheme II was overall destabilization with good stability. The elongation length of the cantilever end determines the load capacity of the two scaffolding schemes as well as the difference in the modes of instability. Compared with that in Scheme I, the elongation of the cantilever end in Scheme II is greatly reduced. The final damage form of the scaffolding changes from local instability to overall instability, and the stability of the formwork improves.

4. Stability Analysis of Socket-Type Keyway Steel Pipe Scaffolding

4.1. *Introduction of the Theory of Second-Order Bending Moment Effects*

The instability failure of steel structures can be divided into two categories [23,24]: equilibrium bifurcation instability and limiting point instability. The instability problem of the steel scaffolding studied in this paper is limiting point instability. The upright rod is subjected mainly to pressure and bending moments (including initial and second-order moments). The initial bending moments of the structure are mainly due to the initial geometric defects of the structure or the bending moments due to lateral loads. Second-order bending moments are those produced by the $P-\delta$ and $P-\Delta$ effects. The $P-\delta$ effect refers to the axial force applied by a rod under the action of flexural deformation to produce an additional bending moment effect. The $P-\Delta$ effect is the effect of the additional bending moment produced by the horizontal deformation of the bar. In the stability calculation of steel scaffolding, the influence of second-order effects is added to the analyzed factors [25]. Lateral-free stiffeners are generally considered for the $P-\delta$ effect. The influence of the $P-\Delta$ effect also needs to be considered in a rigid frame with lateral displacement.

4.2. *Introduction of the Semirigid Node Theory*

Connection nodes are generally categorized into three forms: rigid, articulated, and semirigid connections [26–28]. The upright rods and horizontal rods of the steel scaffolding will rotate relatively during the force process. The connecting nodes of the scaffolding are neither rigid nor articulated. Rather, they are typically semirigid nodes [29,30].

Much research has been carried out on semirigid nodes by previous authors [31]. The rotational stiffness [32] of the semirigid nodes is the most significant factor affecting the overall stress performance of the structure. The rotational stiffness of a semirigid node is related to several factors. The rotational stiffness of a semirigid node can be calculated by the following method:

(1) Experimental methods

The moment-displacement curves of the nodes are obtained by measuring the rotational stiffness of the nodes in the tests. The experimental method allows a more realistic determination of the rotational stiffness of the nodes. The experimental method requires statistical analysis through a large amount of experimental data.

(2) Finite element method

The finite element method is a method applicable to a wide range of problems that can be solved within the field of engineering. Numerical simulation of the connection performance of the nodes is carried out using engineering software. The effect of different factors on the structure can be considered. This approach can be further applied to nonlinear problems [33,34].

The socket-type keyway steel pipe scaffolding is a type of semirigid connected steel frame with lateral movement, according to theoretical analysis [35–37]. The rotational stiffness of this type of semirigid node was derived as $K = 12.02\ kN \cdot \text{m/rad}$ [22] from a large number of tests carried out by previous authors using an experimental method. In contrast, Yu [38] used a computational model of three-story frame columns with coil springs to simulate the restraining moments of beams on columns and the semi-stiffness of nodes.

5. Finite Element Analysis

5.1. Selection of Finite Element Software

Finite Element Analysis (FEA) is a modern computational method that has been rapidly developed for the analysis of structural mechanics. In the field of building structures, the more commonly used general-purpose finite element software are mainly Ansys, PKPM, Midas, Abaqus and SAP2000. SAP2000 v21 is used in this article for finite element modeling and computation.

SAP2000 is a powerful structural analysis software that integrates static-dynamic analysis, load calculation, linear and nonlinear calculation, and other computational analysis in a single package. It contains the latest techniques for static, dynamic, linear and nonlinear analysis, and the calculation process is easy and the results are accurate. The linear elastic buckling analysis and geometric nonlinear analysis contained in the SAP2000 program matches very well with the computational analysis required by the analytical model of the socket-type keyway steel pipe scaffolding. Therefore, SAP2000 is adopted as the analysis software in this article.

In the finite element modeling of the socket-type keyway steel pipe scaffolding, we need to use mainly point objects and line objects. Point objects appear in the form of nodes, which are the most basic units in a structural system. Constraints, springs, loads, connection properties, etc., can be defined and modified in SAP2000 for specified point objects. In the line object, it is possible to define section types, end constraints and displacements, stiffness corrections, connection properties, concentrated and line loads, and even temperature effects on the line object.

5.2. Modeling

SAP2000 software was used to analyze the modeling formwork for two different erection schemes. The upright rods, horizontal rods, and horizontal diagonal rods of the scaffolding are made of round steel pipes with dimensions of $\Phi 48 \times 3.0$. The vertical diagonal rods are made of $\Phi 32 \times 2.0$ round steel pipes. The steels are Q235 carbon structural steel. The modulus of elasticity of the steel is taken as E = 206 MPa, Poisson's ratio is V = 0.3, and the mass density is 7850 kg/m^3. The upright rods are assumed to be rigidly connected. The connection nodes of the upright rods and horizontal rods are considered semirigid nodes. The rotational stiffness of the node is $K = 12.02\ \text{kN} \cdot \text{m/rad}$. Both horizontal and vertical inclined rods are defined as rods articulated at both ends and only axial forces are considered. The support constraint is assumed to be an articulated support that constrains displacements in the X, Y, and Z directions.

The finite element models of scheme I and scheme II are shown in Figure 11.

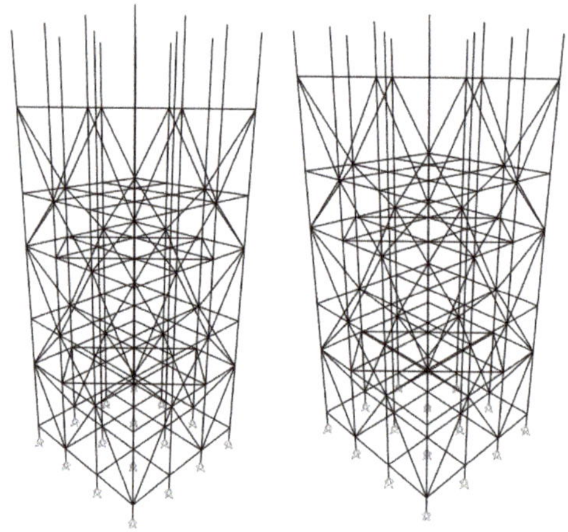

(a) (b)

Figure 11. Finite element model. (**a**) Scheme I. (**b**) Scheme II.

5.3. Linear Elastic Buckling Analysis

Eigenvalue buckling analysis [39] is also known as linear elastic buckling analysis. Mathematically, this process results in a generalized eigenvalue problem, which involves the analysis of first-order linear elastic instability without considering second-order $P-\delta$ (additional effects caused by the deflection of the member under axial pressure) or $P-\Delta$ (additional effects of gravity caused by the horizontal deformation of the member). The eigenvalue equations are solved to determine the ultimate load and damage pattern of the structure when buckling occurs.

Unit loads were applied to the top cantilever end of each upright rod during linear elastic buckling analysis. Buckling was selected for the load type. After the number of flexural modes and the eigenvalue convergence tolerance are defined, the flexural factor is analyzed. The buckling load is determined from the product of the buckling factor and the given unit load. The results of the instability modes for the linear elastic buckling analysis of scheme I and scheme II are shown in Figure 12.

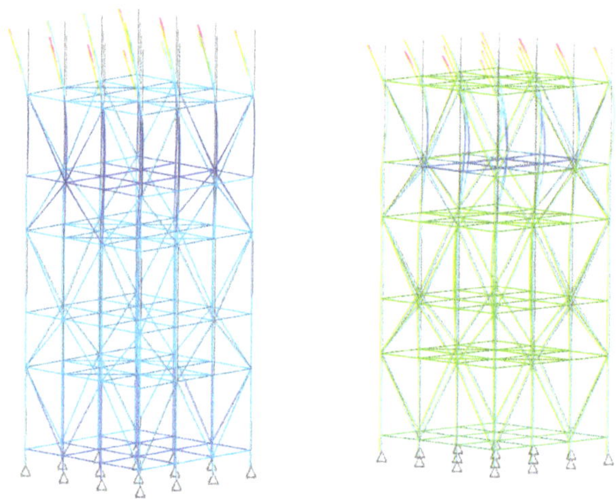

(a) (b)

Figure 12. Finite element analysis model. (**a**) Destabilization mode of scheme I. (**b**) Destabilization mode of scheme II.

The structural instability modes obtained from the finite element linear elastic buckling analysis are consistent with those of the scaffolding tests. The ultimate load capacities of the single upright rods of the whole scaffolding of scheme I and scheme II calculated in SAP2000 were 21.22 kN and 43.14 kN, respectively. The ultimate load capacities obtained from the tests of scheme I and scheme II were 24.50 kN and 34.50 kN, respectively. As shown in Table 3, the calculated results for scheme I are 15% lower than the experimental results. The calculated results for scheme II are 25% greater than the experimental results.

Table 3. Ultimate load capacity from test and finite element analysis.

Scheme	Test	FEA	Errors
I	24.50 kN	21.22 kN	15%
II	34.50 kN	43.14 kN	25%

The finite element analysis results reveal that the damage to the scaffolding in scheme I is a localized instability. The cantilevered end of the top of the upright rod was displaced significantly, and the upper portion of the upright rod was displaced very little. The cantilevered end of the top of the upright rod has a greater elongation length. The ultimate load capacity of the scaffolding is mainly determined by the cantilever end. The damage to the scaffolding in scheme II is an overall instability. The cantilever end of the top of the upright rod was displaced, and the upper portion of the upright rod was also significantly displaced. The ultimate load-carrying capacity does not depend solely on the length of the cantilevered end at the top of the upright rod. The higher ultimate load capacity calculated from the finite element analysis of scheme II than that obtained from the test results occurs because second-order effects and the influence of initial geometric imperfections were not considered in the linear elastic buckling analysis. Therefore, it is necessary to perform a buckling analysis of the scaffolding considering the second-order effects and the initial geometrical defects.

5.4. Geometric Nonlinear Buckling Analysis

The influence of second-order effects should be taken into account when analyzing with finite element software, which is more in line with the force and deformation of the structure under actual working conditions. The over-extension of the cantilever end at the top of the upright rod in scheme I resulted in the ultimate load capacity of the scaffolding being determined by the cantilever end at the top of the upright rod. Localized instability at the cantilever end resulted in the scaffolding not being able to continue to carry the load. The damage to the scaffolding in scheme II is an overall instability. The geometric nonlinear buckling analysis [40] of scheme II considering initial defects was performed by the finite element software SAP2000. The modeling requires attention to the following two points:

(1) The load conditions are defined differently. The buckling analysis case is not chosen for modeling; instead, the static nonlinear loading case is defined. The $P-\Delta$ effect is selected among the geometric nonlinear parameters. The material nonlinear parameters and nonlinear solution control options are set so that the calculation results can converge.

(2) There are three general approaches to nonlinear analysis considering initial geometric defects. The first is the direct simulation of defects, which means that definite initial defects are introduced directly into the finite element model, taking into account the influence of factors such as installation errors and machining processes of the components. The second approach takes into account the effects on the structure due to initial geometric defects by reducing the tangent modulus of the material. The third method is to apply a smaller lateral load in the buckling direction of the structure based on the buckling modal map obtained in the eigenvalue buckling analysis. The most unfavorable effect is obtained by lateral loading. The magnitude of lateral loads

is generally taken as 0.5–0.1% of the vertical force. The third method of analysis is used in this paper; 0.5% of the vertical force is taken as the lateral load value.

As shown in Figure 13, the point at which the maximum displacement occurs is the top point of the cantilever end according to the geometric nonlinear buckling analysis of scheme II. The ultimate capacity of a single upright rod of the structure in the second-order nonlinear buckling analysis is 33.37 kN, which is less than the ultimate capacity in the linear elastic buckling analysis (43.14 kN) but closer to the test result (34.50 kN). The ultimate load capacity of a single upright rod in the second-order nonlinear buckling analysis has only a 4% error from the test results. The nonlinear buckling analysis considering the initial geometric defects is closer to the real force conditions of the structure.

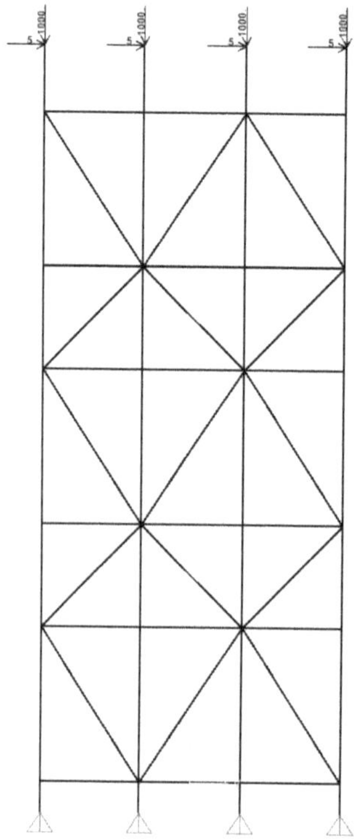

Figure 13. Second-order buckling model for scheme II.

6. Conclusions

(1) For the socket-type keyway steel pipe scaffolding mentioned in this paper, local instability damage occurred at the top of the upright rod when the length of the upright rod at the cantilever end was long (1.2 m). When the length of the cantilever end upright rod is short (0.65 m), the scaffolding is damaged by overall destabilization.

(2) The elongation length of the cantilever end at the top of the scaffolding has a significant effect on the load-bearing capacity. The length of the cantilever end of the scaffolding upright rod is reduced by about 50%, which can increase the ultimate load- carrying capacity of the upright rod by 40%.

(3) The error between the ultimate load capacity in the linear elastic buckling analysis and the experimental results is 25%. In contrast, the error between the calculated ultimate bearing capacity in the nonlinear buckling analysis and the test results was 4%, which is much closer to the test results. Therefore, a buckling analysis of socket-type keyway steel pipes scaffolding was performed. It is recommended to use nonlinear buckling

analysis that takes into account second-order effects and initial geometric defects. This is more in line with the forces on the structure in real situations.

Author Contributions: Conceptualization, C.Z., J.Y., L.J. and Y.H.; Methodology, C.Z.; Validation, C.Z.; Investigation, C.Z.; Resources, J.Y. and L.J.; Data curation, C.Z. and Y.H.; Writing—original draft, C.Z. and Y.H.; Writing—review & editing, C.Z. and Y.H.; Visualization, C.Z. and Y.H.; Supervision, J.Y. and L.J.; Project administration, C.Z. All authors have read and agreed to the published version of the manuscript.

Funding: This research was financially supported by the Hunan Province Engineering Construction Local Standard Making (Revision) Plan Project (Grant numbers: BZ20220005). The sponsor is Hunan Shaping construction Co., Ltd.

Data Availability Statement: Data are contained within the article.

Acknowledgments: The authors wish to acknowledge the National Engineering Laboratory's technical support for High-Speed Railway Construction of the School of Civil Engineering, Central South University. Finally, and most importantly, the authors wish to thank the anonymous reviewers for their thorough evaluations and valuable comments, which have helped improve the paper.

Conflicts of Interest: Author Yanqing He was employed by the company Baoji Works Section of China Railway Xi'an Bureau Group Co., Ltd. The remaining authors declare that the research was conducted in the absence of any commercial or financial relationships that could be construed as a potential conflict of interest.

References

1. Mi, J. Technology Progress of Formwork and Scaffold Industry in China. *Constr. Technol.* **2011**, *40*, 60–63+76.
2. Zhu, Z. Stable Analysis on Fastener Type Full Formwork Support. Master's Thesis, Anhui Jianzhu University, Hefei, China, 2021.
3. Zheng, W. Experimental Study on Integral Stability of Steel Tubular Scaffold with Arch Skeleton Frames. *Fujian Constr. Sci. Technol.* **2021**, *5*, 19–20+51.
4. Sun, Q. Experimental Investigation on the Mechanical Behaviour and Stability of Full Cuplock Steel Tubular Scaffolding. Master's Thesis, Changan University, Xi'an, China, 2018.
5. Huang, H. Application and Safety Supervision of Socket-and-Buckle Scaffold in Construction Engineering. *Jiangsu Build. Mater.* **2022**, 107–109.
6. Xie, Q.; Jin, G.; Wang, Y. Situation of Scaffold in Our Country and Its Development Trends. *Constr. Mech.* **2006**, *27*, 17–21.
7. Wang, S.; Xiao, W.; Cheng, Z. Safety Accident and Prevention of Scaffolding in Construction Site. *Value Eng.* **2023**, *42*, 13–15.
8. Tomasz, N.; Bożena, H. Methodology Based on Causes of Accidents for Forcasting the Effects of Falls from Scaffoldings Using the Construction Industry in Poland as an Example. *Saf. Sci.* **2023**, *157*, 105945.
9. Halperin, K.M.; Michael, M. An Evaluation of Scaffold Safety at Construction Sites. *J. Saf. Res.* **2004**, *35*, 141–150. [CrossRef]
10. Hwang, M.J.; Won, H.J.; Jeong, J.H.; Shin, S.H. Identifying Critical Factors and Trends Leading to Fatal Accidents in Small-Scale Construction Sites in Korea. *Buildings* **2023**, *13*, 2472. [CrossRef]
11. Jia, L.; Liu, H.; Chen, Z. Experimental Research and FEA on Bearing Capacity of Full Hallsteel Tube and Coupler Scaffold Support System. *J. Build. Struct.* **2017**, *38*, 114–122.
12. Ji, M.; Zeng, F.; Dong, Y.; Fan, Y. Calculation Method of Stable Bearing Capacity of Fastener-Type Steel Pipe Formwork Support Upright Rod. *Appl. Sci.* **2023**, *13*, 4838. [CrossRef]
13. Liu, H.; Jia, L.; Wen, S.; Liu, Q.; Wang, G.; Chen, Z. Experimental and Theoretical Studies on the Stability of Steel Tube–Coupler Scaffolds with Different Connection Joints. *Eng. Struct.* **2016**, *106*, 80–95. [CrossRef]
14. Dong, J.; Liu, H.; Lei, M. Overall Structural Stability Analysis on Wheel Coupler Formwork Support. *China Saf. Sci. J.* **2022**, *32*, 85.
15. Yu, H.; Guo, S.; Liu, W.; Gao, R. Experimental Study on Stable Bearing Capacity of Wheel-Buckle Formwork Support Frames. *Structures* **2023**, *48*, 1175–1189. [CrossRef]
16. Chen, Z.; Lv, J.; Liu, H.; Liu, Q.; Ye, B. Experimental and Analytical Studies on Stability of Ultra-Strong Thin-Walled Steel Tube and Coupler Scaffolds. *Eng. Struct.* **2023**, *275*, 115255. [CrossRef]
17. Bian, Y. Experimental Research on Bearing Capacity of New Wheel Buckle Type of Steel Formwork Support. Master's Thesis, Liaoning Technical University, Fuxin, China, 2017.
18. Pieńko, M.; Błazik-Borowa, E. Experimental Studies of Ringlock Scaffolding Joint. *J. Constr. Steel Res.* **2020**, *173*, 106265. [CrossRef]
19. Zhang, L.; Liu, J.; Tang, Q.; Liu, Z. A Numerical Study on Rotational Stiffness Characteristics of the Disk Lock Joint. *J. Constr. Steel Res.* **2023**, *207*, 107968. [CrossRef]
20. Zhang, L.; Lin, B.; Qi, H. Study on the Finite Element Calculation Method of the Bearing Cross Bar of Keyway-Quick Locked Scaffold. In Proceedings of the 8th National Steel Structure Engineering Technology Exchange Meeting, Construction Technology, Xi'an, China, 29 October 2020.

21. Zhang, Z.; Chen, X.; Lin, B.; Cong, J.; Zhang, L. Study on Stability of Casting-Extended Scaffolding Steel Pipe under Axial Compression. *J. Build. Struct.* **2022**, *43*, 228–238.
22. Xie, J. Experimental Research on Mechanical Behavior of Sockettype Steel Tubular Scaffold. Master's Thesis, Central South University, Changsha, China, 2014.
23. Xie, Q. Study on Seismic Design Method Based on P–Δ Effects of Steel Frame. Ph.D. Thesis, Xi'an University of Architecture and Technology, Xi'an, China, 2012.
24. Wang, S.; Zhang, Z.; Wang, C.; Zhu, C.; Ren, Y. Multistep Rocky Slope Stability Analysis Based on Unmanned Aerial Vehicle Photogrammetry. *Environ. Earth Sci.* **2019**, *78*, 260. [CrossRef]
25. Yin, T.; Wang, Z.; Pan, J.; Zheng, K.; Liu, D.; Lu, S. A Design Method for Semi-Rigid Steel Frame via Pre-Established Performance-Based Connection Database. *Buildings* **2022**, *12*, 1634. [CrossRef]
26. Xu, M. The Research on Mechanical Behavior of Steel Frames with Semi-Rigid Connections. Master's Thesis, Changsha University of Science & Technology, Changsha, China, 2005.
27. Li, Y.; He, R. Research for the Semi-Rigid Connection in Frame Structure. *J. Harbin Univ. Civ. Eng.* **1994**, *27*, 112–119.
28. Goverdhan, A. A Collection of Experimental Moment-Rotation Curves and Evaluation of Prediction Equations for Semi-Rigid Connections. Ph.D. Thesis, Vanderbilt University, Nashville, TN, USA, 1983.
29. Chen, W.F.; Kishi, N. Semi-Rigid Steel Beam-to-Column Connection: Data Base and Modeling. *J. Struct. Eng. Struct. ASCE* **1989**, *25*, 105–119. [CrossRef]
30. Xu, L.; Liu, Y. Story Stability of Semi-Braced Steel Frame. *J. Constr. Steel Res.* **2002**, *58*, 467–491. [CrossRef]
31. Yao, H.; Huang, Y.; Ma, W.; Liang, L.; Zhao, Y. Dynamic Analysis of a Large Deployable Space Truss Structure Considering Semi-Rigid Joints. *Aerospace* **2023**, *10*, 821. [CrossRef]
32. Nawar, M.T.; Matar, E.B.; Maaly, H.M.; Alaaser, A.G.; El-Zohairy, A. Assessment of Rotational Stiffness for Metallic Hinged Base Plates under Axial Loads and Moments. *Buildings* **2021**, *11*, 368. [CrossRef]
33. Rui, C. Study and Design of the Static Behavior of Steel Truss Joints of Transmission Towers. Ph.D. Thesis, Chongqing University, Chongqing, China, 2010.
34. Cao, G. Experimental and Theory Study On Unstiffened and Typical Stiffened Steel Tubular Joints of Large-Span-Weight Structures. Ph.D. Thesis, Tongji University, Shanghai, China, 2006.
35. Chen, Z.; Lu, Z.; Wang, X.; Liu, H.; Liu, Q. Experimental and Theoretical Research on Capacity of Unbraced Steel Tubular Formwork Support Based on Sway Frame with Semi-Rigid Connection Theory. *J. Build. Struct.* **2010**, *31*, 11–15.
36. Li, G.; Wang, J.; Liu, Q. A Practical Approach for the Design of Semi-Rigid Composite Frames under Vertical Loads (I)—Design of Beam-to-Column Connections. *Prog. Steel Build. Struct.* **2006**, *6*, 27–37.
37. Zhang, G.; Lv, X.; Liu, J. Displacement-Based Seismic Design of High-Strength Concrete Frame Columns with Confinements. *J. Tongji Univ. Sci.* **2007**, *35*, 143–148.
38. Yu, W.; Wang, Z. Effective Length of Columns in Sway and Sem-rigid Steel Frames. *Build. Sci.* **2008**, *24*, 17–19.
39. Shahin, R.I.; Ahmed, M.; Yehia, S.A. Elastic Buckling of Prismatic Web Plate under Shear with Simply-Supported Boundary Conditions. *Buildings* **2023**, *13*, 2879. [CrossRef]
40. Yao, X.; Yang, J.; Guo, Y. Study on Restoring Force Model of Cold-Formed Thin-Walled Steel Lipped Channel Beam-Columns under Cyclic Load. *Buildings* **2023**, *13*, 114. [CrossRef]

buildings

MDPI

Article

Feasibility Study of Steel Derailment Containment Provisions through Quasi-Static Experiments

Huy Q. Nguyen [1], Hoe-Jin Kim [2], Nam-Hyoung Lim [3], Yun-Suk Kang [4] and Jung J. Kim [1,*]

1 Department of Civil Engineering, Kyungnam University, Changwon 51767, Republic of Korea; nguyenquochuy@muce.edu.vn
2 UB E&C Co., Ltd., 42 Beolmal-ro 50 beon-gil, Bundang-gu, Seongnam 13503, Republic of Korea; lalala1968@naver.com
3 Department of Civil Engineering, Chungnam National University, Daejeon 34134, Republic of Korea; nhrim@cnu.ac.kr
4 Advanced Railroad Civil Engineering Division, Korea Railroad Research Institute, 176 Cheoldobangmulgwan-ro, Uiwang 16105, Republic of Korea; yskang@krri.re.kr
* Correspondence: jungkim@kyungnam.ac.kr; Tel.: +82-552-496-421; Fax: +82-505-999-2165

Abstract: Railway derailments present a safety hazard, carrying the potential for severe consequences for both human lives and the economy. Implementing derailment containment provisions (DCPs) near the track centerline is essential for mitigating risks in operating high-speed rail (HSR) while providing significant advantages for the large-scale upgrade of existing railway infrastructure. Therefore, this paper investigated the feasibility of a DCP system made of steel through quasi-static experiments, aiming to enhance safety in HSR operations. Initially, single anchor tests were conducted to assess its capacity to withstand applied loads, prevent the pullout of steel anchors, and avoid the local rotation of the steel frame. Then, full-scale steel DCP systems were manufactured and tested for quasi-static load at different locations, including the mid-anchor, the mid-span, and the end-anchor. The relationship between applied load and displacement, along with the initial stiffness of the DCP specimens, was discussed. The findings revealed that the single anchor can withstand an applied load of up to 197.9 kN. The DCP specimen maintained structural integrity at the 207 kN target load under all load scenarios, showing a maximum displacement of 8.93 mm in the case of applied load at mid-span. Furthermore, the initial stiffness of the DCP systems was 1.77 to 2.55 times greater than that of a single anchor, validating a force-bearing coordination mechanism among neighboring anchors and the substantial impact of the applied load positions on their stiffness.

Keywords: derailment containment provisions; steel DCP; protection facility; derailment tests; post-derailment safety device; load–displacement; initial stiffness

Citation: Nguyen, H.Q.; Kim, H.-J.; Lim, N.-H.; Kang, Y.-S.; Kim, J.J. Feasibility Study of Steel Derailment Containment Provisions through Quasi-Static Experiments. *Buildings* **2024**, *14*, 171. https://doi.org/10.3390/buildings14010171

Academic Editors: Liqiang Jiang, Wei Chen, Chang He, Yi Hu and Qi Cai

Received: 28 November 2023
Revised: 4 January 2024
Accepted: 6 January 2024
Published: 10 January 2024

1. Introduction

High-speed rail (HSR) systems have risen as pioneers in advanced global transportation, crucial not only for enhancing mobility but also for exerting a profound influence on the dynamics of regional development [1–3]. HSR holds the potential to significantly reduce travel time, improve travel efficiency, and facilitate economic and personnel interactions across different regions and cities, playing a crucial role in promoting a sustainable economy [4]. Nevertheless, the heightened speed of rail has also brought about an augmented risk of derailments, thereby diminishing safety levels during operations [5,6]. Derailments represent the most common type of train accident, resulting in potentially catastrophic consequences for heavy loss of human life and property [7–10]. Achieving complete prevention proves unattainable due to unforeseeable factors like human error, variable weather conditions, and natural disasters [11–14]. In the face of growing demand for HSR development, finding solutions to minimize possible risks becomes increasingly essential.

Scholars worldwide have dedicated their efforts to studying post-derailment behavior and restraining lateral movements of derailed trains to minimize the consequences of derailments, resulting in notable achievements. Barbie et al. suggested employing a brake disc and a bogie frame to maintain derailed vehicles close to the track centerline after evaluating the post-derailment behavior of high-speed rail vehicles through a 3D dynamic model [15–17]. Kajitani et al. devised an L-shaped guide to prevent deviation in derailment accidents, and it has been incorporated across the entire Shinkansen bullet trains in Japan [18]. Sunami et al. designed a post-derailment stopper for bogie frames and proposed a 15-degrees-of-freedom vehicle dynamics model to investigate their motion under a derailment [19]. Guo et al. suggested a safety device mounted under the axle box to minimize trailer vehicle deviation in the event of derailments [20]. Wu et al. developed preventive devices to restrict the lateral displacement of derailed vehicles and verified their effectiveness through derailment experiments conducted at low speeds [21,22]. In general, these studies aimed to improve the guidance ability of vehicle component-based substitute guidance mechanisms by increasing the possibility of contact or collision between the vehicle components and the track to keep the derailed train on the railway track. Efforts to keep derailed trains near the track centerline are beneficial for minimizing damage compared to a part of or the whole train running off the rail or completely veering off the railway tracks [23]. However, the devices must be installed for each bogie in individual trains to achieve optimal safety benefits. Attaining the intended safety enhancements for the entire high-speed rail system entails substantial research and installation costs. Accordingly, the overall cost optimization has been neglected because the risk of derailment is typically associated with high-risk areas during severe weather conditions.

In contrast to the research on enhancing safety systems for individual trains, studies on developing rerouting auxiliary systems for railways in derailments are still scarce. This approach proves beneficial in improving the operational safety of large-scale railway upgrades. Nevertheless, designing an entirely new preventive system to ensure the safety of train operations is likely to be expensive and impractical in the short term, given that many countries worldwide already possess extensive rail networks, with a substantial portion being HSR [24–27]. Consequently, developing auxiliary systems for derailment-prone areas, with the capability of seamlessly integrating them into existing rail infrastructure, becomes even more urgently needed and highly feasible [28]. Recently, derailment containment provisions (DCPs) have emerged as a potential solution for HSR to reduce the consequences by redirecting and maintaining derailed trains near the track centerline. Figure 1 illustrates three commonly used concepts of DCPs in railways, namely DCP Type I, II, and III [29]. DCP Type I is positioned within the track gauge and directly interacts with the wheels during derailments, thereby functioning as guard rails [30–32]. Although DCP Type II has a function similar to its counterpart, Type I, it is positioned outside the tracks. DCP Type III is outside the track gauge but is prepared to absorb impact from axles or bogies rather than the wheels. In Korea, DCP Type III is mandated on railway bridges with train speeds exceeding 200 km/h to prevent collisions with the superstructure or falls from the bridge in a derailment [33].

To the best of our knowledge, no comprehensive investigations have been conducted to explore the design load, installation location, and specifications of derailment containment facilities for HSR. Moreover, there is a shortage of specific objective evidence of DCPs to validate their economic efficiency and feasibility [28]. In efforts to prevent the derailment of HSR, researchers have strived to clarify these issues, aiming to pave the way for the application of DCPs in the HSR system. Lim et al. suggested a modeling method for gravel-filled track ballast, simulating a ballast-wheel collision to study structural responses and impact forces from a derailed train [34]. Song et al. presented a theoretical approach to predict impact loads on reinforced concrete (RC) DCP Type I for HSR and proposed a simplified finite element (FE) model to assess dynamic post-derailment behavior [35]. Bae et al. carried out a full-scale train derailment test to analyze the train's post-derailment behavior and evaluate the performance of RC DCP Type III [36]. Bae et al. also analyzed the functionality of DCP Type I by conducting a comprehensive train derailment test, suggesting an approach to estimate impact loads and assess their containment effect according to changes in the center of gravity during a collision [37]. Nevertheless, upgrading existing railway systems with DCP Types II and III requires substantial foundation structures to absorb impact loads, resulting in extended construction periods. In this scenario, DCP Type I provides a promising solution with the advantages of quality construction, economic efficiency, and faster installation using pre-fabricated components.

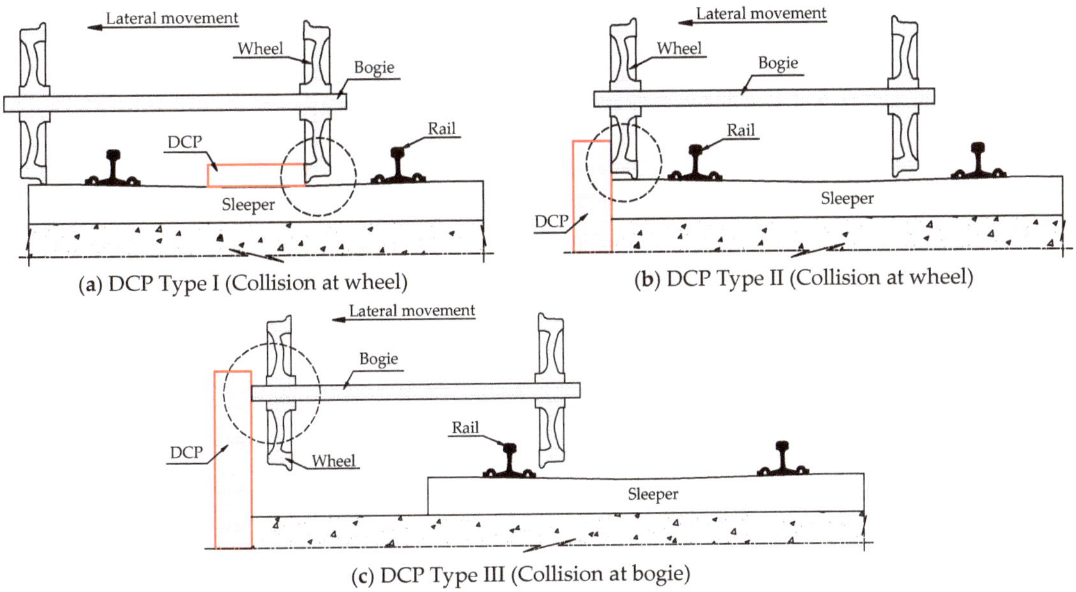

Figure 1. Concepts of DCP in railways.

Additionally, given the importance and ongoing operational requirements of the existing HSR system, DCP Type I may be suitable for meeting the current demand for HSR infrastructure upgrades. While research on DCPs has yielded some achievements, further in-depth assessments are necessary to validate their effectiveness and feasibility throughout experimental tests. Moreover, employing DCPs made of steel offers numerous advantages regarding construction time and deformation compatibility with steel rail systems; however, this area has not received much attention. To fill this gap, a DCP Type I system with steel frames was manufactured and tested under lateral quasi-static loads in this study. The originality of this research is that it included a full-scale experimental test to investigate the response of a steel DCP system Type I under the operational conditions of HSR systems in South Korea or those with comparable requirements, as shown in Figure 1a.

Significance and Scope of the Study

An ongoing project is being carried out to study solutions for Korean HSR to minimize the damage caused by derailment collisions using DCP. Typically, collision/impact tests are conducted to assess both global and local responses in structures, while quasi-static/static tests are mainly employed to reveal global behavior [38]. Predicting the capacity to withstand applied loads corresponding to displacement is crucial to assessing structural safety under impact [39]. Nonetheless, conducting full-scale impact tests for DCP systems at high speeds during derailments is costly and unworkable, preventing the thorough validation of the DCP's load-bearing capacity under adverse working conditions. Hence, the initial phase, which has already been performed, involved assessing the impact forces on the DCP for high-speed trains through collision simulations to propose design specifications for DCPs [40]. As a result, DCP Type I was found to experience a maximum impact load of 165.6 kN in the event of a derailment collision at a speed of 300 km/h on a high-speed rail curve with a radius of 3500 m. In light of these findings, the DCP was designed to assess load-bearing capacity and feasibility through quasi-static tests. In the next stage, these outcomes will play a pivotal role in proposing optimal designs for DCP members, considering factors such as size (length, width, and height), anchor methods, and type of materials, and subsequently in formulating plans for comprehensive impact tests to withstand collision derailment. Finally, an effective post-derailment safety measure for Korean high-speed trains using DCP Type I can be proposed, as shown in Figure 2.

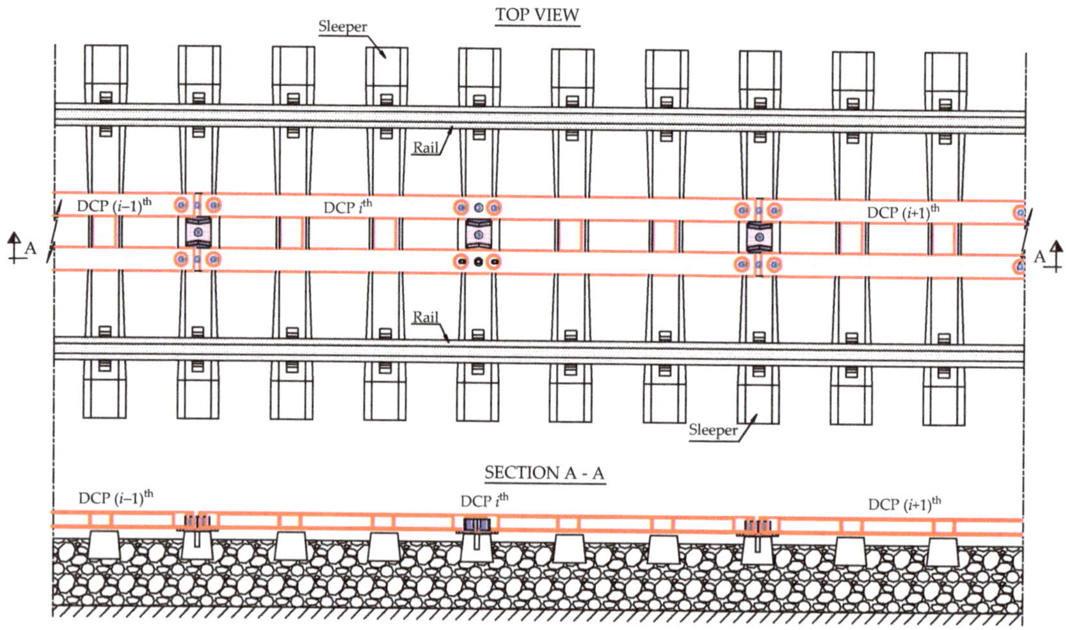

Figure 2. General view of steel frame DCP system Type I.

As we know, steel is widely recognized for its reputation for reliability, which is characterized by consistent and uniform properties. Its appeal is further accentuated by the advantages of enhanced quality control and accelerated erection speed, owing to the precision achieved in factory manufacturing processes. Notably, steel structures emerge as a suitable material for impacted components like DCPs thanks to their flexibility, high ductility, and capacity for impact resistance. Another crucial factor is the deformation compatibility of rails and steel frames under varying temperature conditions, enabling them to operate effectively. This feature gains more importance when considering structures that are integrated in parallel with steel rail DCP Type I systems. The inherent characteristics of steel also contribute to maintaining the structural integrity of the DCP and enhancing its performance under dynamic conditions. As a result, promoting the development of DCP Type I made of rolled steel as a safety measure to prevent the risk of derailments is essential.

Based on the results achieved in the initial stage, the main objective of the proposed experimental study is to evaluate the global response of the steel DCP Type I system under various load location scenarios. The novelty of this study lies in presenting the relationship between the applied load and the displacement and analyzing in detail the influence of the applied load location on the initial stiffness of the DCP Type I system through a full-scale experimental test. In particular, the feasibility of the proposed design under the target load is also discussed and clarified.

2. Experimental Program

2.1. Steel DCP System Details

The steel DCP frame was designed to endure derailment collisions following the guidelines established by Korean researchers, as outlined in the report on the facility development for rail vehicle deviation protection, which was approved by the Ministry of Land, Infrastructure, and Transport of the Korean government [40]. These specimens were employed in experimental investigations to examine their responses under quasi-static loading conditions. The DCP height, proposed through preliminary analysis for high-speed vehicles operated by the Korea Railway Corporation, consists of a 100 mm steel frame, a 20 mm fixed base plate, and a 5 mm insulating rubber sheet. Figure 3a shows the configuration of the steel DCP frame, constructed using assembled modules, with each module measuring 3710 mm in length. The longitudinal beams were crafted from hot-rolled standard rectangular sections measuring 150 × 100 × 9 mm. The steel braces used were of the same section as the longitudinal beam, with a length of 200 mm. Specific dimensions of the fixed base plate, frame fixture, and fixed wedge are referred to in Figure 3c,d. The design compressive strength of the RC sleepers after 28 days was 50 MPa. The yield stress of the steel frame, fixed base plate, frame fixture, and fixed wedge was 355 MPa, with an elastic modulus of 210 GPa.

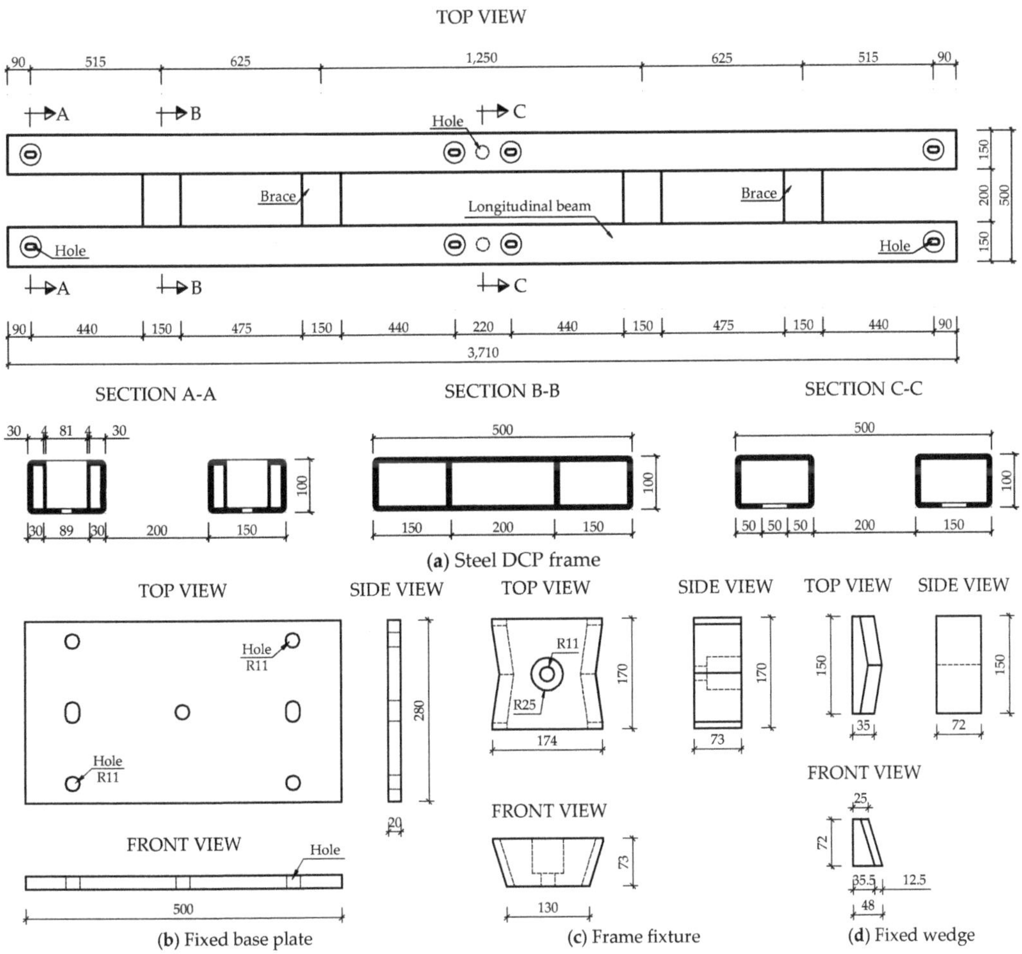

Figure 3. Details about components of the steel DCP system (unit: mm).

Each module of the DCP served the purpose of averting the derailing of a cluster consisting of seven sleepers. The DCP was attached to the sleepers at three positions: the middle anchor and two anchors at both ends. First, for each anchor position, epoxy resin was employed to secure bolts with a 20 mm diameter to the sleepers, each having an anchor length of 90 mm. Next, the sleeper was attached to a fixed base plate and insulation pad using two side fixing nuts. Subsequently, the steel DCP frame was attached to each base plate with four corner bolts. Finally, a frame fixture and fixed wedges were employed to firmly fasten the DCP frame, base plate, and sleeper in place, as shown in Figure 4.

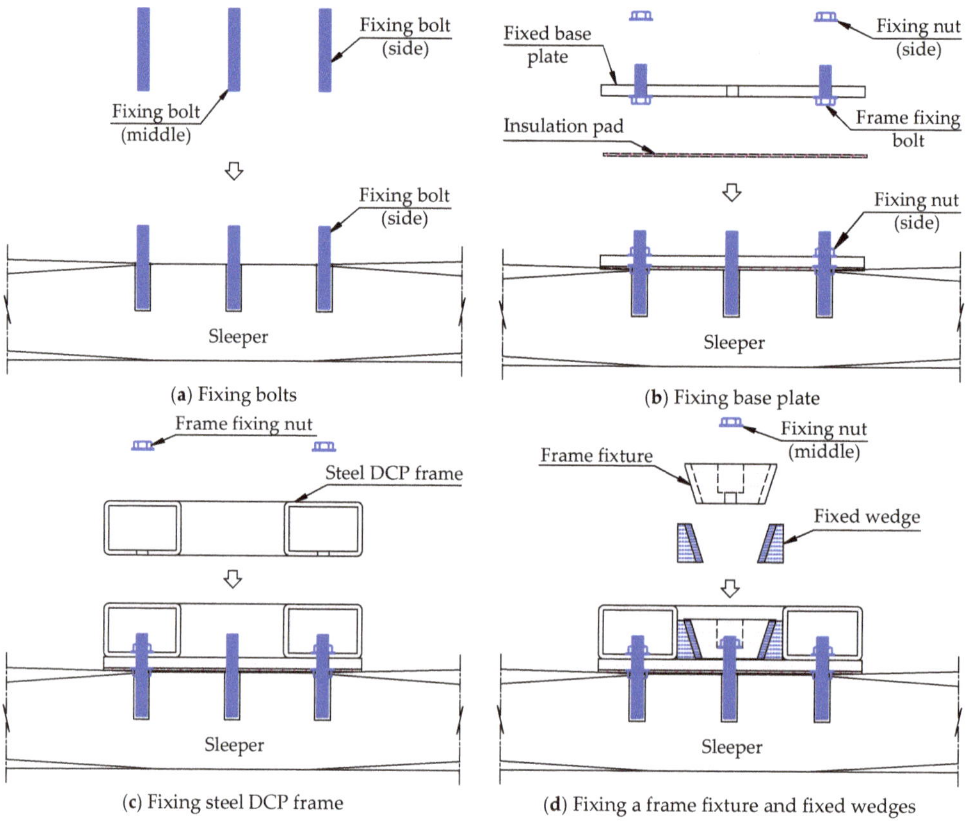

Figure 4. Installation diagram of a steel DCP system.

2.2. Single Anchor Testing

The impact of load on the displacement of single steel anchors was investigated. Figure 5 displays the load applied to the steel DCP frame at the anchor position. The anchor structure was designed similarly to the anchor in the DCP system. Accordingly, it was employed to evaluate the load-bearing capacity of an individual anchor and the corresponding displacement of the tested specimens. We utilized four linear variable displacement transducers (LVDTs) to measure horizontal and vertical displacements. The average of LVDTs H1 and H2 was employed to measure horizontal displacement (H), while LVDTs V1 and V2 were used for vertical displacements (V). Each case was duplicated, with the first and second tests denoted by the suffixes "−1" and "−2", respectively. As shown in Figure 5a, LVDTs with an accuracy level of 0.001 mm were attached around the specimen to measure vertical and horizontal deformations within the gauge length of 100 mm during the single anchor tests, while the load data were measured by the load cell of a universal testing machine (UTM). The tests were conducted at a constant loading rate of 2 mm/min, utilizing a UTM with a capacity of 500 kN.

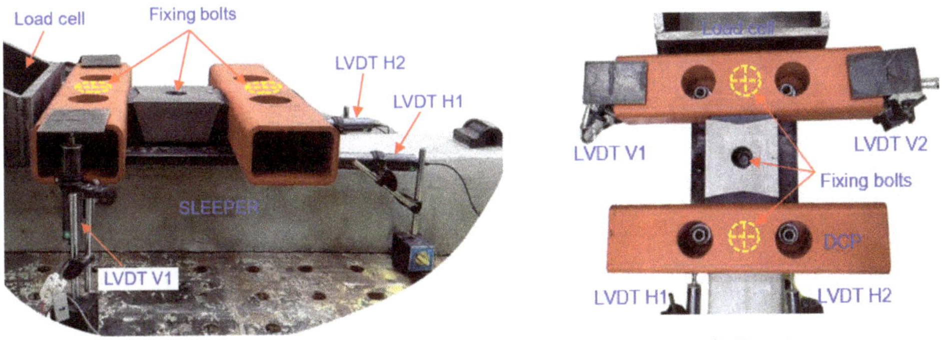

(**a**) Front view (**b**) Top view

Figure 5. Single anchor test setup.

2.3. Steel DCP System Testing

Figure 6 shows the general view of the steel DCP system used in the experimental test, providing a brief overview of its components and design. We divided the steel DCPs into three groups, each designated for quasi-static load testing at different locations: the mid-anchor (referred to as Case 1), the mid-span (Case 2), and the end-anchor (Case 3). We positioned the LVDTs using the load location scenarios corresponding to specific configurations. For Cases 1 and 2, LVDTs L1, L2, and L3 were placed at the end-anchor, mid-span, and mid-anchor, as shown in Figure 7a,b. In Case 3, LVDTs L1 and L2 were placed at the mid-anchor and mid-span, respectively, while LVDTs L3 and L4 were situated on both sides of the end-anchor, as depicted in Figure 7c. In addition, the rigid part employed to secure sleepers serves a role analogous to that of steel rails in a railway system. Each case was repeated twice, with the first and second tests indicated by the suffixes "−1" and "−2", respectively. The sleepers were firmly attached to the rigid part and considered to experience negligible deformation under the effect of load. The testing procedure and equipment specifications for LVDTs and UTM are similar to those used in single anchor tests. The target load was expected to be equal to or higher than the design load (165.6 kN), with a recommended margin exceeding 125%.

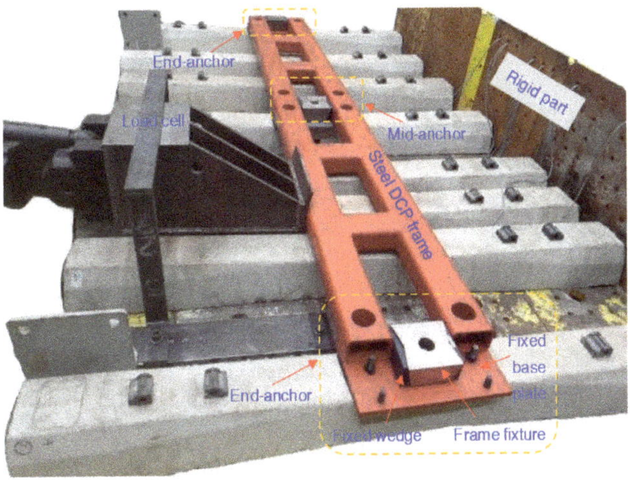

Figure 6. General view of steel DCP test setup.

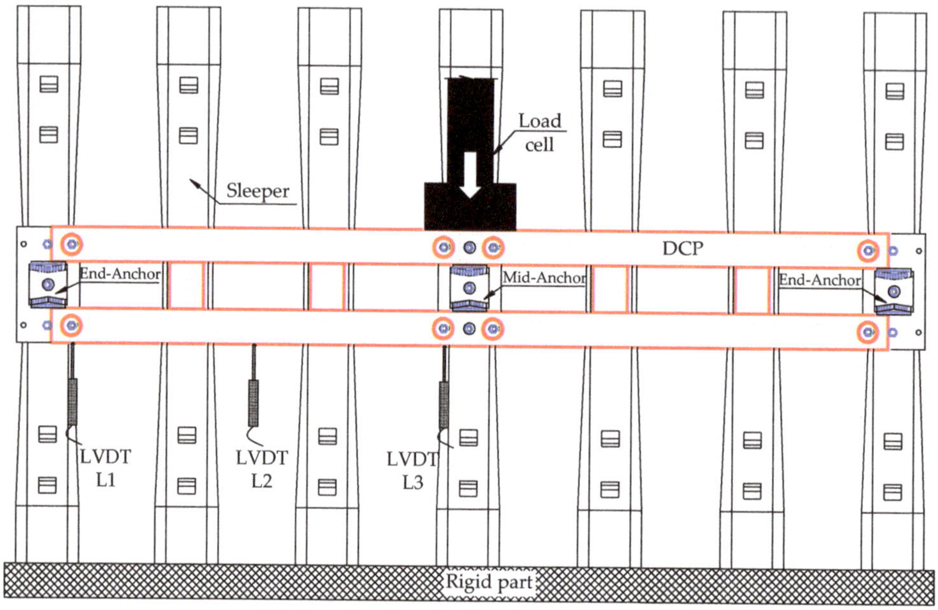

(**a**) Case 1—Applied load at mid-anchor

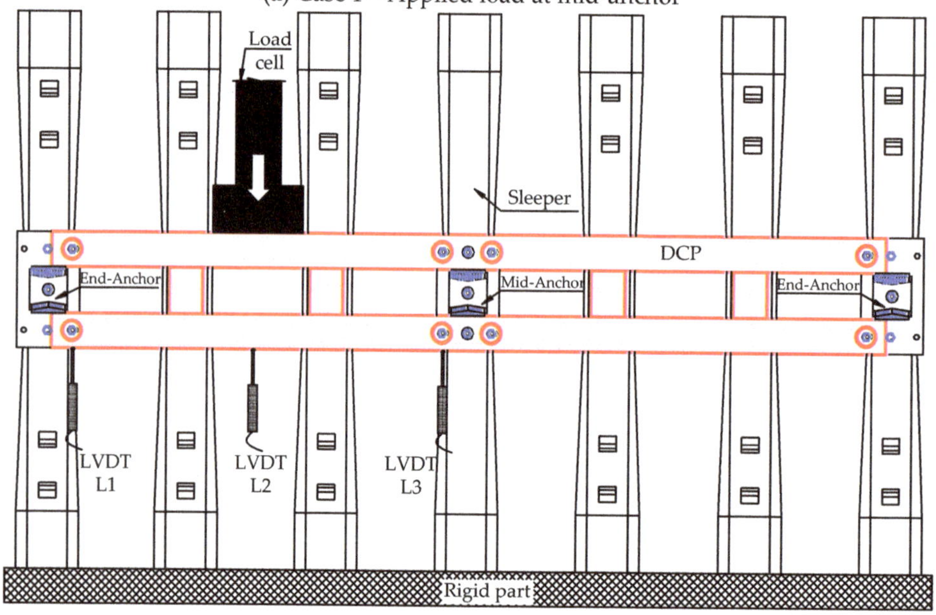

(**b**) Case 2—Applied load at mid-span

Figure 7. *Cont.*

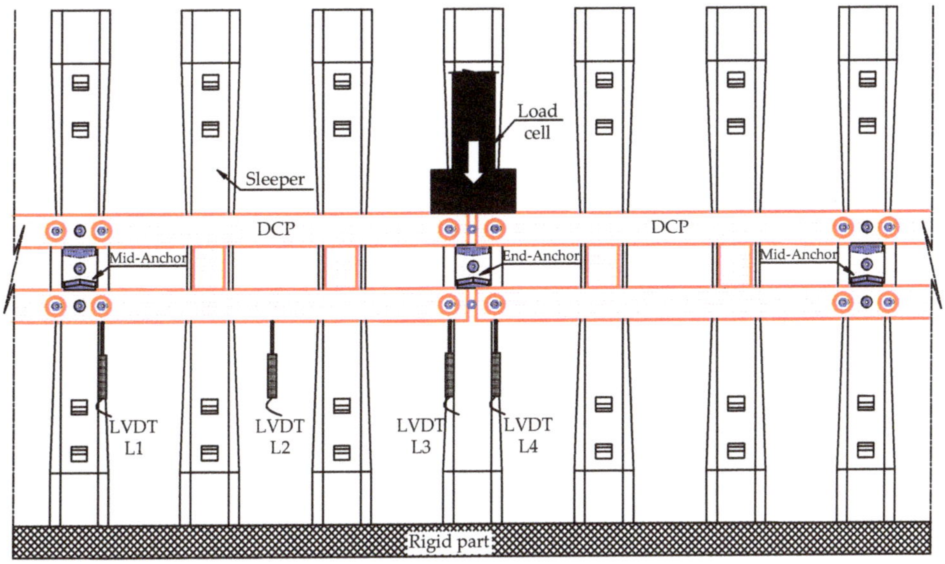

(c) Case 3—Applied load at the end-anchor

Figure 7. Test setup of steel DCPs.

3. Results and Discussion

3.1. Load–Displacement of Single Anchor

This experiment was conducted to assess the performance of the individual anchor during the linear-plastic stage. In Figure 8, the single anchor displayed a linear response until the applied load of 176 kN corresponded to the horizontal displacement of 12.70 mm. Then, the behavior gradually shifted toward the yielding stage with a maximum applied load of 197.9 kN. At this point, the average horizontal and vertical displacements of the single anchor measured 15.77 mm and 0.37 mm, respectively. During the anchor tests, the axial deformation of the high-strength anchor bolt was notably lower than the displacement observed for the remaining components constituting the anchor. The main factor influencing anchor displacement under load was the disparity in diameter between the hole and the anchor bolt, coupled with the deformation of the insulation pad, fixed base plate, and DCP steel frame. Notably, the negligible vertical displacement observed was evidence of the feasibility of the anchor structure, especially the connection between the bolts and sleepers in preventing pullout and local rotation. The results from the tests involving single anchors are presented in Table 1.

Table 1. Experimental data of the single anchor.

Specimen ID.	P_u (kN)	P_y (kN)	Δu_H (mm)	Δu_V (mm)	Δy_H (mm)	K_i (kN/mm)
S1	199.9	176.0	16.34	−0.88	11.95	13.86
S2	195.8	176.0	15.20	0.14	11.86	13.97
Average	197.9 (2.05)	176.0 (0)	15.77 (0.57)	−0.37 (0.51)	11.90 (0.04)	13.92 (0.05)

Notes: P_y and Δy_H represent the yielding load and the associated displacement; P_u, Δu_H, and Δu_V denote the highest applied load and the respective vertical and horizontal displacements; K_i refers to the initial stiffness; the values in parentheses are the standard deviations.

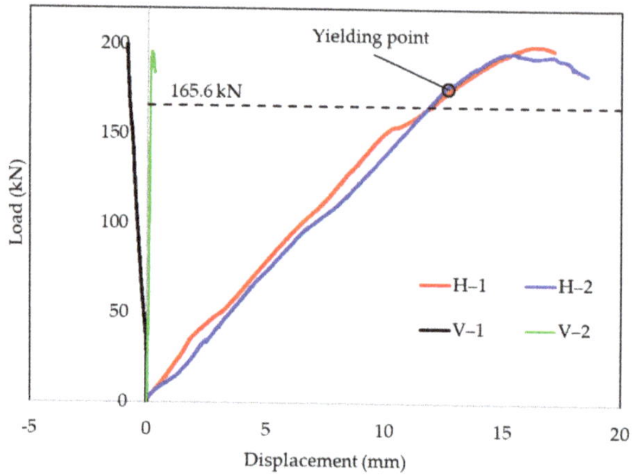

Figure 8. Load–displacement relationship of single anchor.

3.2. Load–Displacement of Steel DCP System

Figure 9 displays the load–displacement curves for the DCP system in the three cases, with the load cell applied at the mid-anchor (Case 1), mid-span (Case 2), and end-anchor (Case 3). Each load case was repeated twice to obtain the average value to ensure accuracy. The full-scale structural tests were successfully conducted, with the recorded data mostly clustering around the average value and exhibiting low standard deviations, as indicated in Table 2. Overall, the DCP system showed a linear response until reaching the target load of 207 kN, which exceeded 125% of the designed load of 165.6 kN. In Figure 9a–c, the observation validated that the structure could uphold its integrity under applied loads with elastic deformations recognized. Notably, the LVDTs' displacement did not return to the original position after unloading, with the primary cause being localized displacements in the components constituting the anchor. For Case 1, the average displacement values of LVDTs L1, L2, and L3 were 4.35 mm, 5.35 mm, and 5.84 mm, respectively. Even though the LVDT at the mid-anchor indicated the maximum value, these displacements showed only a minor discrepancy. This suggested that the main factor contributing to the DCP displacement was the deformation of the components constituting the anchor, given the negligible deflection of LVDTs observed in the steel DCP, as shown in Figure 9a.

Case 2 showed average displacements of 8.93 mm for LVDT L1, 7.83 mm for LVDT L2, and 6.98 mm for LVDT L3, as depicted in Figure 9b. It is worth mentioning that the maximum displacement at the end-anchor surpasses that achieved at the mid-span (the applied load position) due to the significant difference in stiffness between the steel DCP frame and the steel anchor. In Case 3, LVDTs L3 and L4 exhibited the most substantial displacement with an average of 8.43 mm, while LVDTs L1 and L2 recorded displacements of 1.74 mm and 4.64 mm, respectively, as illustrated in Figure 9c. It was evident that under the same applied load (207 kN), the maximum displacement of 8.93 mm in Case 2 was 1.53 and 1.06 times higher than Cases 1 and 3, respectively. It indirectly validated that Case 2 might be the most adverse loading scenario. The experimental data for all tested cases are summarized in Table 2.

To sum up, the proposed design maintained its structural integrity. It exhibited maximum displacements of 4.58 mm in Case 1, 7.16 mm in Case 2, and 6.07 mm in Case 3 at the designed load of 165.6 kN. Moreover, during the single anchor test, the steel bolts reached the yield stage at the applied load of 176 kN, while the DCP system continued to respond with linear elasticity of the applied load of 207 kN or higher. It showed substantial differences in the cooperative forces between neighboring anchors and the redistribution of internal forces within the DCP systems in cases of different applied loads. Based on the design criteria, the proposed DCP system demonstrated feasibility for its implementation within the Korean railway system. It is worth mentioning that the present analysis only covers the most unfavorable loading scenarios without considering safety-biased group effects among DCPs.

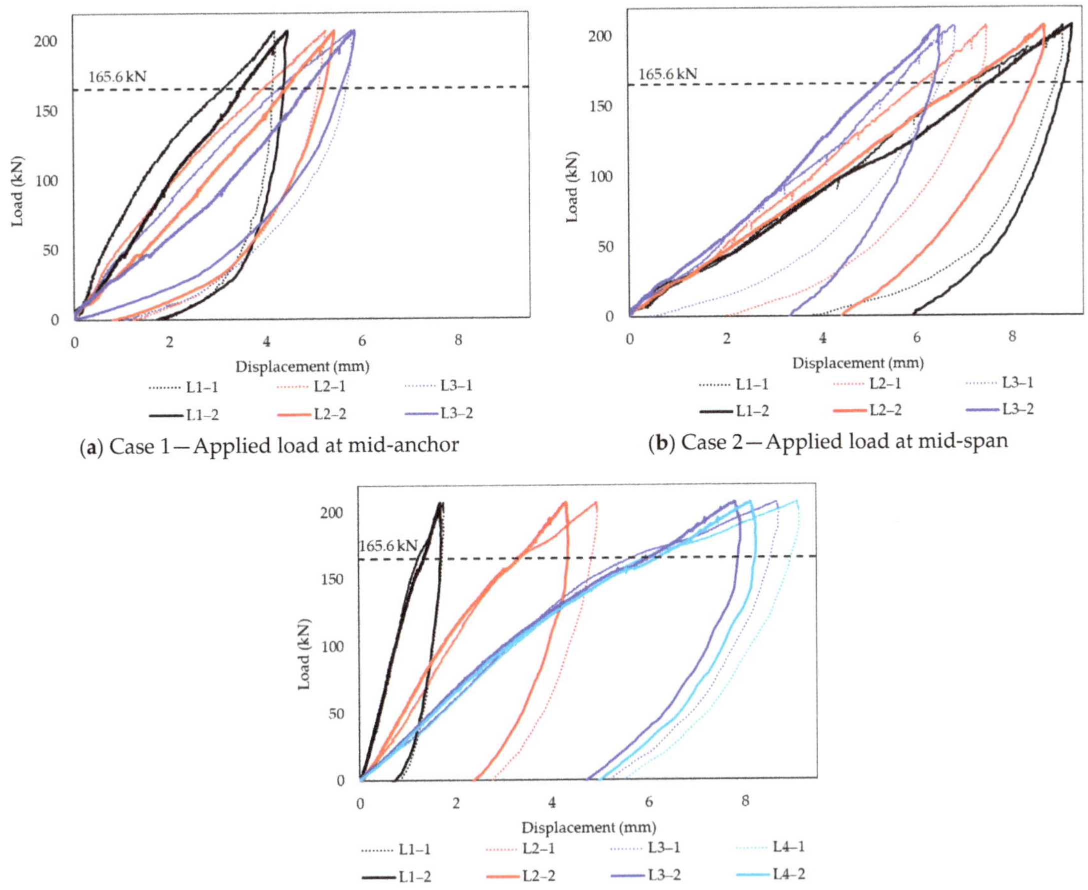

(**a**) Case 1—Applied load at mid-anchor

(**b**) Case 2—Applied load at mid-span

(**c**) Case 3—Applied load at end-anchor

Figure 9. Load–displacement relationship of DCP with different load locations.

Table 2. Experimental data of the steel DCP system.

Notation	Case 1—Load at Mid-Anchor			Case 2—Load at Mid-Span			Case 3—Load at End-Anchor		
	L1	L2	L3	L1	L2	L3	L1	L2	Aver. L3&L4
Designed load (165.6 kN)	3.32 (0.19)	4.20 (0.24)	4.58 (0.29)	7.16 (0.36)	6.32 (0.52)	5.41 (0.35)	1.28 (0.06)	3.34 (0.00)	5.95 (0.17)
Target load (207 kN)	4.35 (0.13)	5.35 (0.10)	5.84 (0.03)	8.93 (0.35)	7.83 (0.88)	6.98 (0.49)	1.74 (0.04)	4.64 (0.32)	4.35 (0.46)
Initial stiffness (kN/mm)	35.44 (0.20)			26.78 (3.00)			24.62 (1.33)		

Notes: the values in parentheses are standard deviations.

3.3. Initial Stiffness Analysis

The initial stiffness of the structures could affect the displacement and the displacement ductility estimation in displacement-based designs, which mainly depend on factors like shape, size, material properties, and support conditions [41,42]. The slope of the load–displacement curve during the linear elastic stage presents the initial stiffness of the DCP system. It is determined by dividing the applied load by the respective displacement at the load application point [43]. In this study, Case 1 showed the highest initial stiffness among the tested cases, estimated at 35.44 kN/mm. On the other hand, the initial stiffness in Case 2 (26.78 kN/mm) and Case 3 (24.62 kN/mm) was relatively lower, approximately 1.32 to 1.44 times less than that observed in Case 1. These results demonstrated that the initial stiffness of the steel DCP system was significantly affected by the positions of the applied loads. In addition, we could confirm the mobilization of the bearing capacity between neighboring anchors in the proposed structure when the initial stiffness in Cases 1, 2, and 3 exceeded the initial stiffness of the single anchor (13.92 kN/mm) by factors of 2.55, 1.92, and 1.77, respectively. This analysis provided greater insight into the structural responses, serving as a cornerstone for advancing simulation research and addressing the multifaceted challenges associated with mitigating damage from a train derailment.

4. Conclusions

This study investigated the behavior of DCP systems made of steel under different quasi-static loading scenarios. The primary focus was on analyzing the global response of the proposed systems to evaluate their feasibility in mitigating damage during train derailments. The study also concerned the relationship between the applied load and displacement, coupled with the initial stiffness of the DCP systems. The conclusions drawn from the study are as follows:

The anchor structure proved its capability to withstand impact forces by effectively preventing the pullout and local rotation of the steel frame through the single anchor test, achieving a yield strength of 176 kN and a maximum load of 197.9 kN.

The DCP specimens revealed maximum displacements of 7.16 mm at 165.6 kN while maintaining elasticity and structural integrity at 207 kN. The main factor causing DCP displacement was the deformation and localized displacements of the components constituting the anchor, with the maximum displacement in Case 2 measuring 8.93 mm, which exceeded Case 1 by 1.53 times and Case 3 by 1.06 times.

The initial stiffness of the DCP systems, ranging from 1.77 to 2.55, exceeded that of the single anchor. This verified the force-bearing coordination mechanisms among neighboring anchors and the notable differences between loading scenarios.

The proposed system would be viable for minimizing damage during derailments. Nevertheless, further research on impact loads, local bearing capacity, and the reliability of the DCP system is recommended to meet the specified standards in Korea and comparable railway systems.

Author Contributions: Conceptualization, J.J.K.; methodology, H.-J.K., N.-H.L. and J.J.K.; software, H.Q.N. and Y.-S.K.; validation, H.-J.K., N.-H.L. and Y.-S.K.; formal analysis, H.Q.N. and Y.-S.K.; investigation, H.-J.K., N.-H.L. and J.J.K.; data curation, J.J.K. and H.Q.N.; writing—original draft preparation, H.Q.N. and J.J.K.; writing—review and editing, H.Q.N., Y.-S.K., H.-J.K., N.-H.L. and J.J.K.; visualization, H.-J.K. and N.-H.L.; supervision, J.J.K.; project administration, H.-J.K. and N.-H.L.; funding acquisition, N.-H.L. All authors have read and agreed to the published version of the manuscript.

Funding: This work was supported by a Korea Agency for Infrastructure Technology Advancement (KAIA) grant funded by the Ministry of Land, Infrastructure and Transport (Grant RS-2021-KA163289).

Data Availability Statement: Data are contained within the article.

Conflicts of Interest: Author Hoe-Jin Kim was employed by the company UB E&C Co., Ltd. The remaining authors declare that the research was conducted in the absence of any commercial or financial relationships that could be construed as a potential conflict of interest.

References

1. Kim, H.; Sultana, S.; Weber, J. A geographic assessment of the economic development impact of Korean high-speed rail stations. *Transp. Policy* **2018**, *66*, 127–137.
2. Jia, S.; Zhou, C.; Qin, C. No difference in effect of high-speed rail on regional economic growth based on match effect perspective? *Transp. Res. Part A Policy Pract.* **2017**, *106*, 144–157. [CrossRef]
3. Verma, A.; Sudhira, H.S.; Rathi, S.; King, R.; Dash, N. Sustainable urbanization using high speed rail (HSR) in Karnataka, India. *Res. Transp. Econ.* **2013**, *38*, 67–77.
4. Yang, X.; Lin, S.; Zhang, J.; He, M. Does High-Speed Rail Promote Enterprises Productivity? Evidence from China. *J. Adv. Transp.* **2019**, *2019*, 1279489.
5. Yanhui, L.; Al-Bukhaiti, K.; Shichun, Z.; Abas, H.; Nan, X.; Lang, Y.; Yu, Y.X.; Daguang, H. Numerical study on existing RC circular section members under unequal impact collision. *Sci. Rep.* **2022**, *12*, 14793. [PubMed]
6. Montenegro, P.A.; Carvalho, H.; Ribeiro, D.; Calçada, R.; Tokunaga, M.; Tanabe, M.; Zhai, W.M. Assessment of train running safety on bridges: A literature review. *Eng. Struct.* **2021**, *241*, 112425.
7. Cao, Y.; An, Y.; Su, S.; Xie, G.; Sun, Y. A statistical study of railway safety in China and Japan 1990–2020. *Accid. Anal. Prev.* **2022**, *175*, 106764.
8. Evans, A.W. Fatal train accidents on Europe's railways: An update to 2019. *Accid. Anal. Prev.* **2021**, *158*, 106182.
9. Liu, X.; Rapik Saat, M.; Barkan, C.P.L. Freight-train derailment rates for railroad safety and risk analysis. *Accid. Anal. Prev.* **2017**, *98*, 1–9.
10. European Union Agency for Railways. *Report on Railway Safety and Interoperability in the EU*; Publications Office of the European Union: Luxembourg, 2022.
11. Zhang, Z.; Turla, T.; Liu, X. Analysis of human-factor-caused freight train accidents in the United States. *J. Transp. Saf. Secur.* **2021**, *13*, 1157–1186.
12. Dindar, S.; Kaewunruen, S.; An, M. A hierarchical Bayesian-based model for hazard analysis of climate effect on failures of railway turnout components. *Reliab. Eng. Syst. Saf.* **2022**, *218*, 108130. [CrossRef]
13. Wu, X.; Chi, M.; Gao, H.; Ke, X.; Zeng, J.; Wu, P.; Zhu, M. Post-derailment dynamic behavior of railway vehicles travelling on a railway bridge during an earthquake. *Proc. Inst. Mech. Eng. Part F J. Rail Rapid Transit.* **2014**, *230*, 418–439. [CrossRef]
14. Heleno, R.; Montenegro, P.A.; Carvalho, H.; Ribeiro, D.; Calcada, R.; Baker, C.J. Influence of the railway vehicle properties in the running safety against crosswinds. *J. Wind Eng. Ind. Aerodyn.* **2021**, *217*, 104732. [CrossRef]
15. Brabie, D.; Andersson, E. Dynamic simulation of derailments and its consequences. *Veh. Syst. Dyn.* **2006**, *44* (Suppl. S1), 652–662. [CrossRef]
16. Brabie, D. Wheel-Sleeper Impact Model in Rail Vehicles Analysis. *J. Syst. Des. Dyn.* **2007**, *1*, 468–480. [CrossRef]
17. Brabie, D.; Andersson, E. Post-derailment dynamic simulation of rail vehicles–methodology and applications. *Veh. Syst. Dyn.* **2008**, *46* (Suppl. S1), 289–300. [CrossRef]
18. Kajitani, Y.; Kato, H.; Asano, K. Development of an L-shaped guide to prevent deviation from rails. *JR-EAST Tech. Rev.* **2010**, *15*, 53–56.
19. Sunami, H.; Morimura, T.; Terumichi, Y.; Adachi, M. Model for analysis of bogie frame motion under derailment conditions based on full-scale running tests. *Multibody Syst. Dyn.* **2012**, *27*, 321–349. [CrossRef]
20. Guo, L.; Wang, K.; Lin, J.; Zhang, B.; Chen, Z.; Song, X.; Du, G. Study of the post-derailment safety measures on low-speed derailment tests. *Veh. Syst. Dyn.* **2016**, *54*, 943–962. [CrossRef]
21. Wu, X.; Chi, M.; Gao, H.; Zhang, D.; Zeng, J.; Wu, P.; Zhu, M. The study of post-derailment measures to limit the extent of a derailment. *Proc. Inst. Mech. Eng. Part F J. Rail Rapid Transit.* **2014**, *230*, 64–76. [CrossRef]
22. Wu, X.; Chi, M.; Gao, H. The study of post-derailment dynamic behavior of railway vehicle based on running tests. *Eng. Fail. Anal.* **2014**, *44*, 382–399. [CrossRef]
23. Tang, Z.; Hu, Y.; Wang, S.; Ling, L.; Zhang, J.; Wang, K. Train post-derailment behaviours and containment methods: A review. *Railw. Eng. Sci.* **2023**. [CrossRef]
24. Zhang, D.; Peng, Y.; Xu, Y.; Du, C.; Zhang, Y.; Wang, N.; Chong, Y.; Wang, H.; Wu, D.; Liu, J.; et al. A high-speed railway network dataset from train operation records and weather data. *Sci. Data* **2022**, *9*, 244. [CrossRef] [PubMed]
25. Büchel, B.; Spanninger, T.; Corman, F. Empirical dynamics of railway delay propagation identified during the large-scale Rastatt disruption. *Sci. Rep.* **2020**, *10*, 18584. [CrossRef] [PubMed]
26. Cheng, Y.-s.; Loo, B.P.Y.; Vickerman, R. High-speed rail networks, economic integration and regional specialisation in China and Europe. *Travel Behav. Soc.* **2015**, *2*, 1–14. [CrossRef]
27. Guo, B.; Ke, J. The Impacts of High-speed Rail on Sustainable Economic Development: Evidence from the Central Part of China. *Sustainability* **2020**, *12*, 2410. [CrossRef]
28. Bae, H.-U.; Yun, K.-M.; Moon, J.; Lim, N.-H. Impact Force Evaluation of the Derailment Containment Wall for High-Speed Train through a Collision Simulation. *Adv. Civ. Eng.* **2018**, *2018*, 2626905. [CrossRef]

29. Hamilton, B.A. *Report on the Findings of: Current Practice and Effectiveness of Derailment Containment Provisions on High Speed Lines*; HSL-Zuid Organisation: Zoetermeer, The Netherlands, 2004.
30. Lai, J.; Xu, J.; Wang, P.; Chen, J.; Fang, J.; Ma, D.; Chen, R. Numerical investigation on the dynamic behaviour of derailed railway vehicles protected by guard rail. *Veh. Syst. Dyn.* **2021**, *59*, 1803–1824. [CrossRef]
31. Mao, J.h.; Xiang, J.; Gong, K. Mechanism and Application of a New Guard Rail for Improving the Stability of Small Radius Curve Tracks with Continuous Welded Rails. In Proceedings of the 2013 Fourth International Conference on Digital Manufacturing & Automation, Qindao, China, 29–30 June 2013; pp. 773–779.
32. Nishimura, K.; Terumichi, Y.; Morimura, T.; Adachi, M.; Morishita, Y.; Miwa, M. Using Full Scale Experiments to Verify a Simulation Used to Analyze the Safety of Rail Vehicles During Large Earthquakes. *J. Comput. Nonlinear Dyn.* **2015**, *10*, 031013. [CrossRef]
33. Authority, K.R.N. *Railway Design Guideline and Handbook—Subsidiary and Safety Facilities for Main Lined*; KR C-02060; Korea Rail Network Authority: Daejeon, Republic of Korea, 2017.
34. Lim, N.-H.; Kim, K.-J.; Bae, H.-U.; Kim, S. DEM Analysis of Track Ballast for Track Ballast–Wheel Interaction Simulation. *Appl. Sci.* **2020**, *10*, 2717. [CrossRef]
35. Song, I.-H.; Koo, J.-S.; Shim, J.-S.; Bae, H.-U.; Lim, N.-H. Theoretical Prediction of Impact Force Acting on Derailment Containment Provisions (DCPs). *Appl. Sci.* **2023**, *13*, 3899. [CrossRef]
36. Bae, H.-U.; Moon, J.; Lim, S.-J.; Park, J.-C.; Lim, N.-H. Full-Scale Train Derailment Testing and Analysis of Post-Derailment Behavior of Casting Bogie. *Appl. Sci.* **2020**, *10*, 59. [CrossRef]
37. Bae, H.-U.; Kim, K.-J.; Park, S.-Y.; Han, J.-J.; Park, J.-C.; Lim, N.-H. Functionality Analysis of Derailment Containment Provisions through Full-Scale Testing-I: Collision Load and Change in the Center of Gravity. *Appl. Sci.* **2022**, *12*, 11297. [CrossRef]
38. Pham, T.M.; Hao, H. Influence of global stiffness and equivalent model on prediction of impact response of RC beams. *Int. J. Impact Eng.* **2018**, *113*, 88–97. [CrossRef]
39. Fujikake, K.; Li, B.; Soeun, S. Impact Response of Reinforced Concrete Beam and Its Analytical Evaluation. *J. Struct. Eng.* **2009**, *135*, 938–950. [CrossRef]
40. Chungnam-National-University; Korea-Railroad-Research-Institute; CS-Global; UBI-E&C; Korea-Railroad-Authority-&-Road-Kinematics. *Facility Development and Performance Standard Research for Rail Vehicle Deviation Protection*; Chungnam National University: Daejeon, Republic of Korea, 2020.
41. Tena-Colunga, A. Aspects to Consider in the Assessment of Effective Stiffness for Reinforced Concrete Beams. *J. Arch. Eng.* **2021**, *27*, 04020048. [CrossRef]
42. Ding, F.; Ding, H.; He, C.; Wang, L.; Lyu, F. Method for flexural stiffness of steel-concrete composite beams based on stiffness combination coefficients. *Comput. Concr.* **2022**, *29*, 127–144.
43. Duy, N.P.; Hiep, D.V.; Anh, V.N. Performance of concrete beams reinforced with various ratios of hybrid GFRP/steel bars. *Civ. Eng. J.* **2020**, *6*, 1652–1669.

buildings

Article

Seismic Performance Evaluation and Retrofit Strategy of Overhead Gas-Insulated Transmission Lines

Xiaoxuan Li [1], Qiang Xie [1] and Jiayi Wen [2,*]

[1] College of Civil Engineering, Tongji University, Shanghai 200092, China; 1610225@tongji.edu.cn (X.L.); qxie@tongji.edu.cn (Q.X.)
[2] College of Civil Engineering, Nanjing Forestry University, Nanjing 210037, China
* Correspondence: njfuwjy@njfu.edu.cn; Tel.: +86-13817913451

Abstract: The overhead gas-insulated transmission line (GIL) in ultra-high-voltage converter stations, distinct from traditional buried pipelines, demands a thorough investigation into its seismic behavior due to limitations in existing codes. A refined finite element model is established, considering internal structure, slip between various parts, and the relative displacement at the internal conductor joint. Seismic analysis reveals the vulnerability of the GIL at the corner of the pipeline height change, with two failure modes: housing strength failure and internal conductor displacement exceeding the limit. Furthermore, the acceleration amplification coefficient of the support generally exceeds 2.0. Two retrofit methods, namely increasing the fundamental frequency of all supports and fixing the connections between all supports and the housing, have been proposed. The results indicate the effectiveness of both methods in reducing the relative displacement. Fixing all the supports effectively reduces the stress, whereas the other one yields the opposite effect. The seismic performance of a GIL is determined not by the dynamic amplification of supports, but by the control of relative displacement between critical sections, specifically influenced by the angular deformation of the pipeline's first-order translational vibration mode along the line direction. Seismic vulnerability analysis reveals a reduction of over 50% in the failure probability of the GIL after the retrofit compared to before the retrofit, with the PGA exceeding 0.4 g.

Keywords: gas-insulated transmission line (GIL); seismic performance; dynamic amplification; seismic vulnerability; seismic retrofit

Citation: Li, X.; Xie, Q.; Wen, J. Seismic Performance Evaluation and Retrofit Strategy of Overhead Gas-Insulated Transmission Lines. *Buildings* **2023**, *13*, 2968. https://doi.org/10.3390/buildings13122968

Academic Editor: Gianfranco De Matteis

Received: 5 November 2023
Revised: 20 November 2023
Accepted: 25 November 2023
Published: 28 November 2023

1. Introduction

A gas-insulated transmission line (GIL) is a crucial component within an ±800 kV converter power station, serving as a central connectivity hub that reliably links various functional zones, as depicted in Figure 1. In the figure, the I and II denote the bipolar configuration within the ultra-high voltage converter station, where bipolar operation in ultra-high voltage direct current transmission involves utilizing two DC circuits—termed positive and negative circuits—to transmit electrical energy effectively in the system. In ultra-high-voltage substations, the GIL is typically designed as an overhead pipeline to minimize space occupation and the need for extensive civil engineering work. This compact layout, combined with its substantial transmission capacity and high reliability, positions the GIL as an ideal choice for long-distance electric power transmission [1,2].

Shown in Figure 2, a GIL embodies an overhead pipeline structure that utilizes three-pillar insulators (TPIs) to establish a coaxial relationship between the internal conductor and the outer housing. The TPI not only ensures uniform distribution of the internal magnetic field, but also enables the GIL standard straight-line segment length to reach 18 m, significantly reducing installation complexity and costs. Both the outer housing and the internal conductor are constructed from aluminum, while the TPI is crafted from cast resin. The space between the outer housing and the internal conductor is filled with insulating gas, fulfilling the engineering criteria of significantly reducing equipment size and spacing.

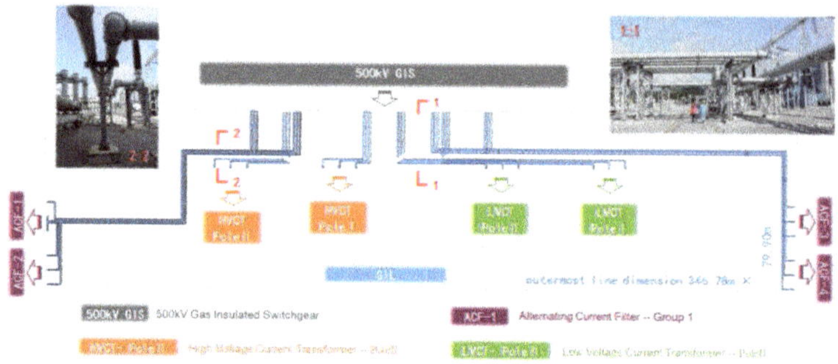

Figure 1. Gas-insulated transmission line.

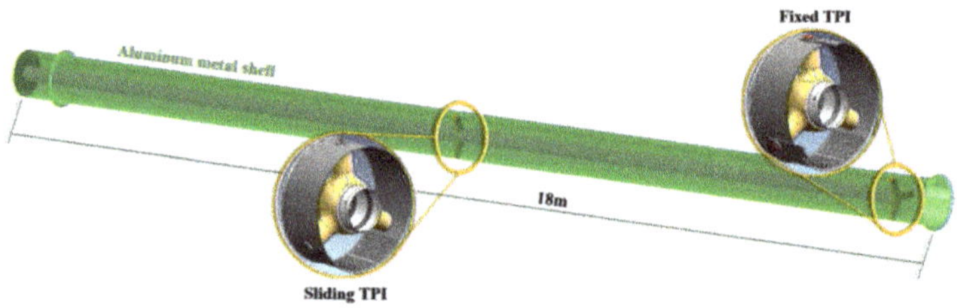

Figure 2. Standard linear unit structural form of GIL.

In the face of burgeoning electricity demands and a growing reliance on long-distance GIL transmission in ultra-high-voltage converter power stations, particularly within regions prone to frequent seismic activity [3–6], it has become increasingly imperative to gain profound insights into the seismic behavior of GIL systems. At present, seismic research on pipelines mostly focuses on water [7–9], oil [10,11], and gas [12–14] pipelines, and most of them are buried pipelines [15–19]. The other part is for indoor small-diameter water pipes or non-structural pipes [20–22], which are mostly suspended pipes. Jahangiri et al. [15] assess economic seismic risks for buried steel gas pipelines, finding factors like thickness ratio, burial depth-to-diameter ratio, soil properties' impact failure probabilities, and economic losses. Karamanos et al. [7] investigate seismic methodologies for buried steel water pipelines, emphasizing soil interaction modeling, resistance, mitigation measures, and design provisions. Sultanov et al. [18] address seismic effects on underground pipelines, proposing a wave theory-based model that highlights the significance of dynamic stress considerations. Obviously, seismic research for buried pipelines should prioritize soil conditions and pipeline–soil interaction. In the case of the GIL, an overhead system comprising supports, pipelines, and internal conductors, the emphasis is on investigating the dynamic amplification effect of supports, seismic responses of the pipeline itself, and the interior of the pipeline.

This article presents an advanced finite element model that faithfully replicates diverse connections between support, housing, and internal conductors to thoroughly investigate the seismic behavior of GILs. The study initiates by pinpointing the critical seismic response and weak positions of GILs through nonlinear time history analysis. Following this, three retrofit strategies are introduced, accompanied by an in-depth analysis of the key factors influencing GILs' seismic performances. Ultimately, the article employs seismic vulnerability analysis to evaluate the failure probability of a GIL before and after effective retrofit at different Peak Ground Accelerations (PGAs).

2. Seismic Performance of GIL

To evaluate the seismic performance of a complex overhead pipeline structure such as a GIL, it is necessary to first clarify the structural characteristics of its various components and the different connection methods between them, and then use dynamic characteristic analysis to grasp its main vibration modes and natural frequency range. Finally, we select appropriate seismic motion records and perform seismic response calculations on the GIL to explore its key seismic responses and weak locations.

2.1. The Finite Element Model of GIL

This paper focuses on the seismic performance of two specific GILs, namely the incoming GIL of the first group of 550 kV AC filters (ACF-1) and the second group of AC filters (ACF-2), as shown in Figure 3. The line lengths of the GILs in two horizontal directions are 117.5 m and 78.3 m, respectively. The heights of the internal conductor axes of each section of the ACF-1 incoming GIL relative to the ground are 5.405, 2.200, 12.500, 7.670, and 4.925 m, respectively. Notably, the layout of the ACF-2 incoming line is closely coincident with that of the ACF-1 incoming line, with a mere 1 m variation in height. It is worth noting that both GILs traverse through the entrance of a factory building, following a plane layout, and are elevated using a steel platform. As a result, the maximum height difference between the GILs reaches 10.3 m, potentially rendering them susceptible to seismic vulnerability.

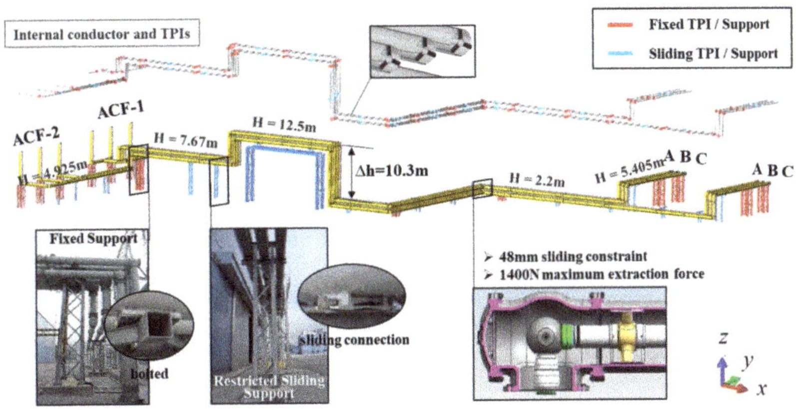

Figure 3. The composition and structure of GIL.

From a structural perspective, a gas-insulated transmission line (GIL) constitutes a complex and large-span system, encompassing supports, pipelines, and internal conductors. The critical structural components responsible for force transmission include the inner conductor, three-pillar insulators (TPIs), outer housing, and supports. The finite element model for GIL components utilizes shell elements for the housing due to their intricate forces, beam elements for the support, TPIs, and an inner conductor based on their structural characteristics. The finite element model of the GIL system is depicted in Figure 3. To provide a comprehensive understanding, Table 1 outlines the materials, cross-sectional dimensions, and main mechanical properties of each component.

Table 1. Structural and mechanical parameters of GIL components.

Components	Dimensions (mm)	Elastic Modulus (GPa)	Density (kg/m^3)	Ultimate Strength (MPa)
Inner conductor	Φ 175.0, t 7.5 [1]	69	4225	220
TPI [2]	Φ 41.1	15	2320	30
Outer housing	Φ 258.0, t 8.0 [1]	69	2690	220
Supports	/	206	7850	235

Φ [1] represents the outer diameter, t represents the thickness; TPI [2] with solid cross-section.

As illustrated in Figure 3, TPIs and supports play crucial roles as supporting structures for the inner conductor and outer housing, respectively. To account for boundary constraints, they can be categorized into two types: fixed supports/TPIs and limited sliding supports/TPIs. For both types of supports, their bottom ends are anchored to the ground. The top of the fixed support is firmly connected to the housing through bolts, effectively limiting the movement of the housing in all directions at the connection point, while the top of the sliding support is connected to the housing through a limit device, which limits all degrees of freedom of the housing except for the axis. Along the axis direction, the housing is only subjected to friction. Therefore, fixed supports employ coupling to simulate support–housing connections, ensuring dynamic convergence by associating a control point on the support with the controlled surface of the housing. The connection between the sliding support and the housing is modeled using the connector element SLOT, resembling a one-way moving pin. This element facilitates controlled points on the housing's directional movement, allowing the friction coefficient to be set at 0.3. The movement of the sliding support and the housing in other directions is interlinked.

For two types of TPI, which share the same main structure, they differ solely in the method of connection with the housing. As depicted in Figure 2, the central component of a TPI is a hollow aluminum sleeve, cast together with three epoxy resin insulators that are evenly spaced along the circumference of the sleeve. The internal conductor passes through this hollow aluminum sleeve and is welded to the TPI. Consequently, all TPIs and the internal conductor form an integrated unit, with their relative positions remaining constant throughout their operation. The distinguishing feature between the two TPI types lies in their respective connection methods with the housing. For the fixed TPI, direct welding with three metal connectors secures it to the housing. In contrast, the sliding TPI employs nylon rollers for connection, facilitating limited sliding movement along the axis direction within the inner side of the housing. For the connection between fixed/sliding three-pillar insulators (TPIs) and the housing, a coupling is uniformly used for simulation. The points on the housing are selected as the control points, and the points on the base of the TPI are selected as the controlled points. The difference is that the coupling connection between the sliding TPI and the housing will release the axial degrees of freedom and couple the remaining degrees of freedom. The fixed TPI is coupled to the surgical coupling in all six directions of freedom.

In addition, considering the influence of thermal expansion, a displacement margin of 48 mm with a maximum extraction force of 1400 N is left at the corner joint of the internal conductor of the GIL, as shown in Figure 3. The inner conductor joint is simulated using the connector element SLOT, illustrating its axial mechanical behavior, shown in Figure 4.

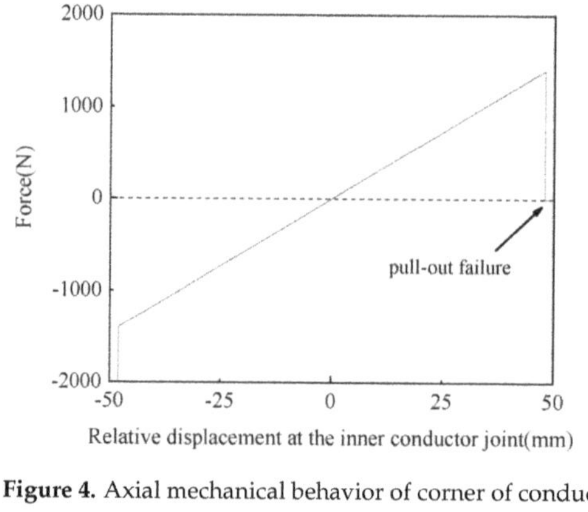

Figure 4. Axial mechanical behavior of corner of conductor.

The nonlinearity in our model is primarily manifested in two aspects: the friction between the sliding support and the housing, and the connection at the corner of the inner conductor.

2.2. Modal Analysis of GIL and Selected Ground Motions

The modal analysis results of the GIL model indicate that its frequency range of the first 120 orders is 1.99 to 9.80 Hz, indicating that it belongs to a modal-intensive structure. And the natural frequency range of the GIL coincides with the predominant frequency range of earthquakes, which is not conducive to earthquake resistance. The first six modal shapes of the GIL model are shown in Figure 5 and can be mainly divided into two categories: one is the vibration of the sliding support along the pipeline axis, and the other is the local deformation of the pipeline accompanied by the support. In fact, this is a common feature of all mode shapes in GILs. In the direction of the pipeline axis, the sliding support provides little constraint on the pipeline, and at places with large height differences or corners, the pipeline deformation is obvious. In addition, the vibration frequencies of the sliding support and the pipeline are not the same, and further analysis is needed for the dynamic amplification effect of the support on the pipeline; the two may generate significant relative displacement in earthquakes.

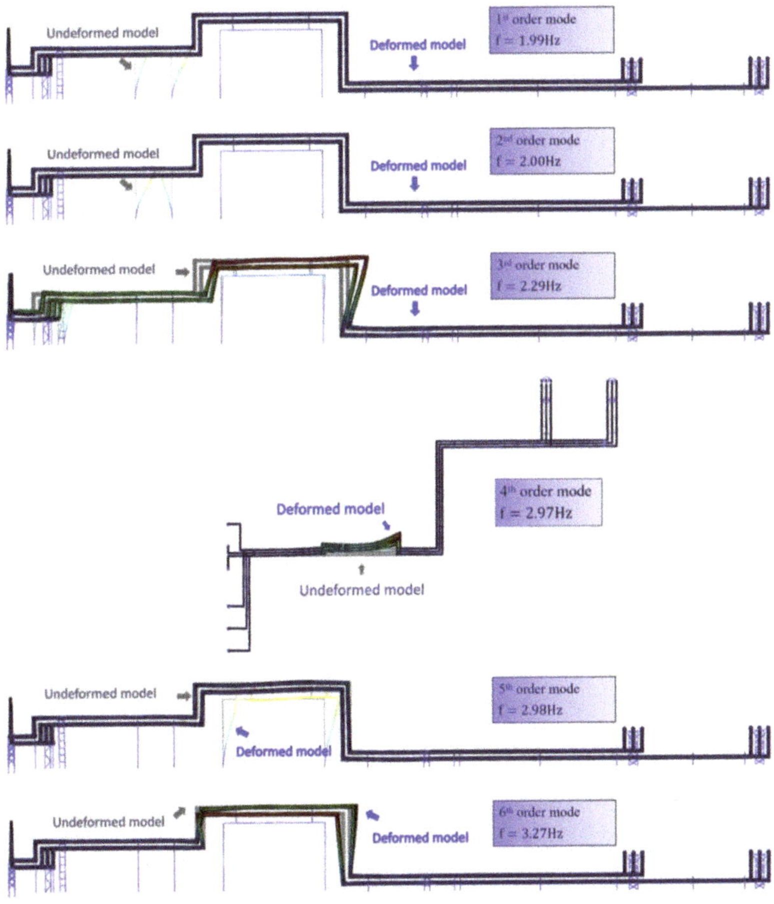

Figure 5. The first six vibration modes of the GIL model.

According to the Chinese regulation Code for Seismic Design of Electrical Installations (GB 50260-2013) [23], seven sets of seismic ground motion records were selected for seismic response analysis, including five sets of natural earthquake records and two sets of artificial waves, as listed in Table 2.

Table 2. Information of seven sets of selected ground motions.

Record	Earthquake	Station	Year	Mw	Epicentral Distance (km)
El Centro	Imperial Valley—02	El Centro Array #9	1940	6.95	12.98
Chi-Chi	Chi-Chi, Taiwan	TCU054	1999	7.62	37.64
RSN66	San Fernando, CA, USA	Hemet Fire Station	1971	6.61	153.8
RSN82	San Fernando, CA, USA	Port Hueneme	1971	6.61	80.21
RSN1527	Chi-Chi, Taiwan	TCU100	1999	7.62	42.77
Arti-1 (RSN1101)	Kobe, Japan	Amagasaki	1995	6.90	38.79
Arti-2 (RSN4031)	San Simeon, CA, USA	Templeton	2003	6.50	36.63

The acceleration response spectra of the selected seismic records and the average response spectrum of seven sets of seismic records are plotted in accordance with the required response spectrum (RRS), as shown in Figure 6. It can be seen that in all three directions, within the frequency range corresponding to the main vibration modes of the structure, the average response spectrum can better cover the RRS. When calculating seismic response, the ground peak acceleration ratio in the three directions of XYZ is 1:0.85:0.65. According to the importance level of the substation to which the GIL belongs and the seismic fortification requirements of the site [24], and in accordance with the requirements of the Code for Seismic Design of Electrical Installations (GB 50260-2013) [22], the peak ground acceleration in the X-direction is classified as a nine-degree seismic fortification intensity and a rare occurrence intensity (seismic fortification intensity), with a probability of exceeding it in 50 years of 10%; rare earthquakes with a 50-year probability of exceeding 2% to 3% are taken as 0.4 g and 0.62 g, respectively.

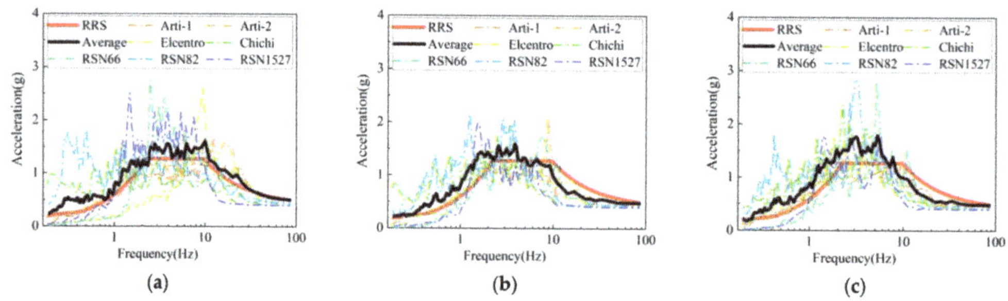

Figure 6. RRS, acceleration response spectra and average acceleration response spectrum of selected ground motions. (**a**) X-direction; (**b**) Y-direction; (**c**) Z-direction.

2.3. Key Seismic Responses and Vulnerable Locations of GIL

The seismic performance of GILs can be evaluated from two aspects: stress response and relative displacement response. Figure 7a,b respectively list the stress response peaks and relative displacement peaks at some key locations in a GIL under seven sets of ground motions.

Regarding stress response, the main focus is on whether the peak stress responses of the TPI, internal conductor, and housing exceed their corresponding ultimate stress. According to the information provided by the design manufacturer, the ultimate stress of the TPI is 30 MPa. Even when the PGA is 0.62 g, both the average peak stress response and the maximum peak stress response under seven sets of seismic motions do not exceed the ultimate stress of the TPIs. The stress safety factor is defined as the ratio of the ultimate stress to the average peak stress response under seismic action. It can be obtained that when the PGAs are 0.4 g and 0.62 g, the stress safety factors of the three-pillar insulators are 3.26 and 1.69, respectively, which comply with the requirement that the stress safety factor should be greater than 1.67 in GB 50260-2013. Therefore, it can be considered that the three-pillar insulators will not undergo stress damage under earthquake action. The ultimate stress of both the internal conductor and the housing is 220 MPa. From Figure 7a, it can be seen that the peak stress response of the internal conductor is much smaller than the ultimate stress, and there is no risk of stress failure. However, for the peak stress response of the housing, when the PGA is 0.4 g, although the average peak stress does not exceed the limit stress, the stress safety factor is only 1.55, which does not meet the specification requirements. When the PGA is 0.62 g, the average peak stress under the seven seismic actions exceeds the limit stress. Therefore, from the perspective of stress response, the possible failure mode of the GIL is housing stress failure.

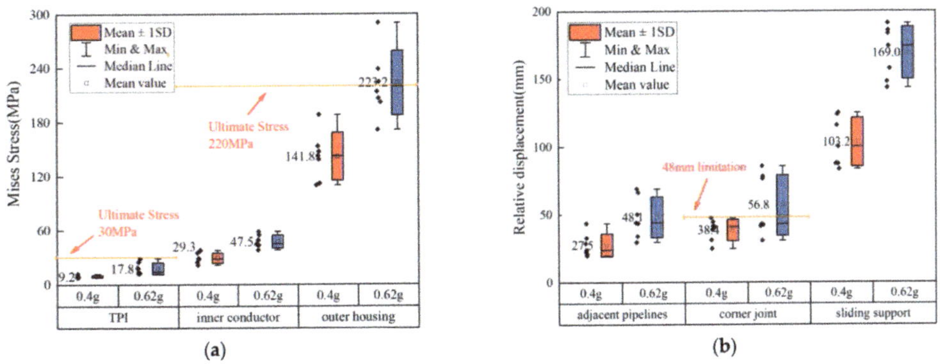

Figure 7. Key seismic response peaks at different parts of a GIL. (**a**) Peak stress responses; (**b**) peak relative displacement responses.

For displacement response, the main focus is on the peak relative displacement response between adjacent pipelines (A, B, C three-phase), internal conductor corner joints, and sliding supports and housings. The distance between adjacent pipelines is 338 mm. From Figure 7b, it can be seen that the relative displacement between adjacent pipelines under seismic action is much smaller than the distance, so there is no risk of collision between them. With respect to the relative displacement at the corner joint of the internal conductor, as previously described in Figure 3, it should be less than 48 mm. However, when the PGA is 0.62 g, the average peak relative displacement at the corner joint of the internal conductor will exceed the limit displacement by nearly 20%. This can cause damage to the connector by pulling it out, affecting the normal operation of the GIL's electrical functions and causing significant losses. There is no limit to the relative displacement between the sliding support and the housing, but under the condition of a 0.4 g earthquake, the average peak relative displacement will reach 103.2 mm. When the PGA is 0.62 g, the average peak relative displacement will reach 169.0 mm. Therefore, from the perspective of displacement response, the possible failure mode of the GIL is internal conductor corner pull-out failure.

Figure 8 shows the weak position of a GIL under earthquake action, which is the position where the maximum peak stress response of the housing occurs or the position where the maximum relative displacement occurs at the corner joint of the internal conductor. It can be seen that the maximum seismic response of the GIL is concentrated at the vertical corner of the line, and its position does not change with the increase in PGA.

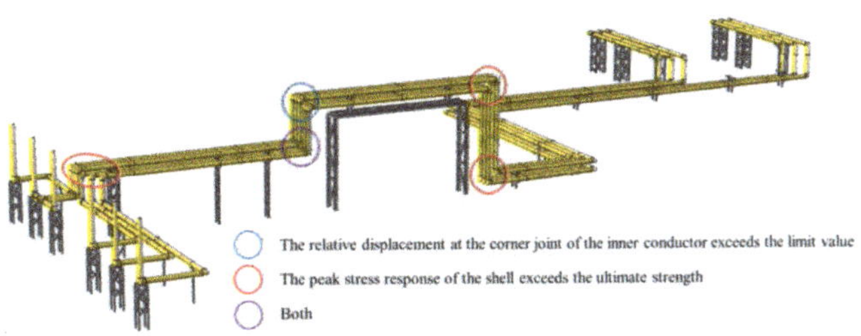

Figure 8. The weak position and failure mode of GIL under earthquake.

3. Retrofit Measures of GIL

According to relevant Chinese regulations [25,26], the seismic design of overhead pipelines is often simplified into a single-degree-of-freedom system for seismic response calculation, or through the implementation of structural measures to ensure seismic reliability. However, findings from the preceding section reveal the seismic vulnerability of GILs at specific corners where there is a change in pipeline height, indicating a potential composite failure mode involving housing strength and the risk of excessive relative displacement of the inner conductor. Consequently, the standardized calculation method offers only limited insight into the peak displacement and maximum stress response of the support, falling short of the comprehensive seismic performance evaluation demanded by the intricate GIL system. This underscores the need for an in-depth exploration of the dynamic response characteristics inherent in the GIL, a complex support–pipeline system, with the aim of developing precise seismic enhancement strategies.

3.1. The Dynamic Amplification Effect of Supports

The initial focus of analysis pertains to the dynamic amplification effect of supports. As detailed in Section 2.1, GIL supports are classified into fixed and sliding types. For clarity in subsequent analyses, each support type is systematically designated and numbered within the GIL system, as depicted in Figure 9. Fixed supports are predominantly located at specific points: the initial section (connected to a GIS), the terminal section (adjacent to the AC filter bushing), and the line corners, representing a comparatively smaller subset. Conversely, sliding supports are evenly distributed along straight line segments, averaging 13.05 m per piece. Figure 10 provides a visual representation of the configuration and heights of all supports. Fixed supports adopt a four-column steel structure, forming composite spatial architectures ranging in height from 1.95 m to 7.40 m. In contrast, sliding supports utilize a two-column steel structure, spanning heights from 1.95 m to 12.05 m.

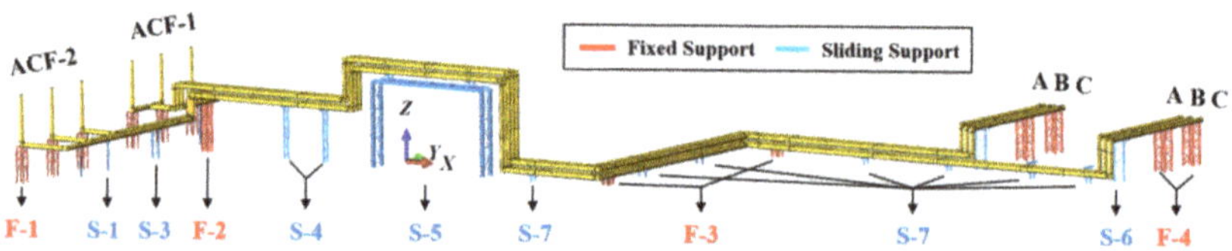

Figure 9. Layout and numbering of support types.

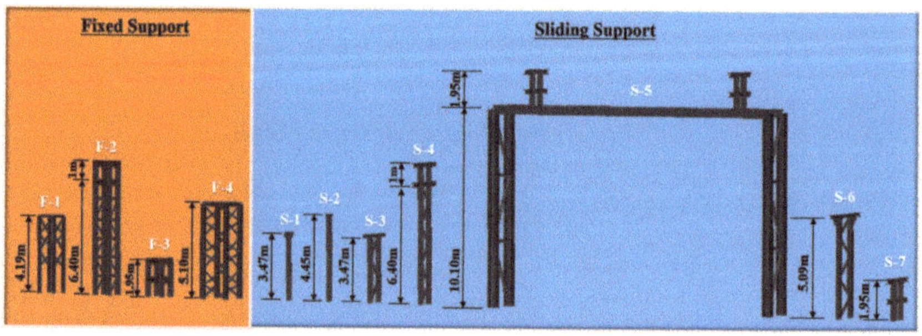

Figure 10. Outline drawings and numbering of different supports.

Based on the findings in Section 2, the vulnerable locations of a GIL are primarily situated in the latter segment. Consequently, this section focuses on the supports and pipelines in the latter half of a GIL. Figure 11a illustrates the acceleration amplification coefficients at the tops of various supports and the corresponding housing linked to the support along the pipeline axis direction under seven sets of ground motions; "H" denotes the housing, and "S" signifies the support. The acceleration amplification coefficient is defined as the ratio of the peak acceleration response to the peak ground acceleration.

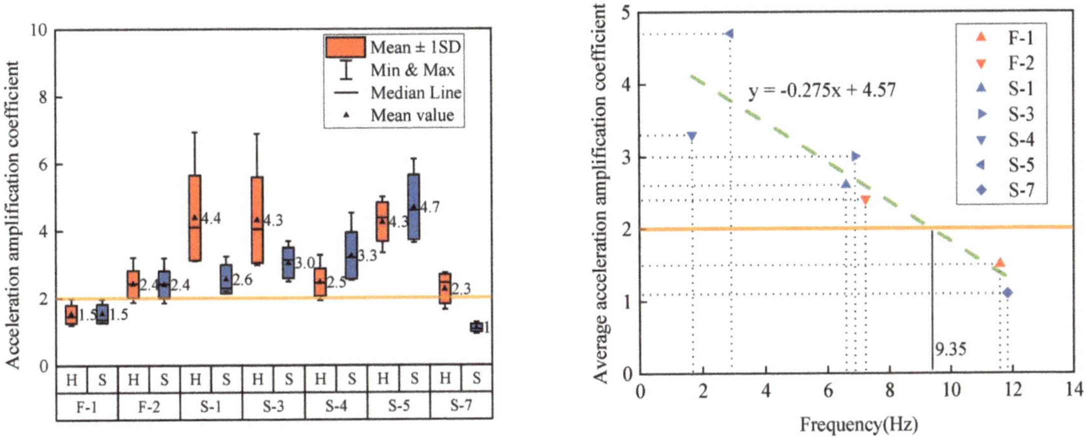

(**a**) (**b**)

Figure 11. Support dynamic amplification effect. (**a**) Acceleration amplification coefficient of supports and housing; (**b**) relationship of support fundamental frequency and amplification coefficient.

Figure 11a reveals several crucial findings. Firstly, with the exceptions of F-1 and S-7, the average acceleration amplification coefficients for all other support types and the housing consistently surpass 2.0. Secondly, a considerable variation exists in the average acceleration amplification coefficients across different support types, ranging from 1.1 to 4.7. Similarly, there are significant differences in the average acceleration amplification coefficients of the corresponding housing with different supports, ranging from 1.5 to 4.4. Thirdly, the presence of asynchronous vibrations along the axis between the housing and the sliding support leads to a notable disparity in acceleration amplification coefficients, with a ratio spanning from 0.76 to 2.09. This diversity corresponds to two distinct scenarios: for S-1, S-3, and S-7, the housing's acceleration amplification coefficient significantly exceeds that of the support, whereas for S-4 and S-5, an inverse trend is observed. These outcomes emphasize that the seismic performance of a GIL is inadequately evaluated by simplifying this complex system as a single-degree-of-freedom structure and conducting seismic response calculations.

Figure 11b illustrates the scatter plot and linear fitting relationship between the natural frequency of the support and the acceleration amplification coefficient. According to the GIL analysis presented in this study, restraining the dynamic amplification coefficient of the support to less than 2.0 requires a fundamental frequency exceeding 9.35 Hz. Moreover, to ensure that the support does not amplify the ground input, the fundamental frequency should be greater than 12.98 Hz. However, in many instances, the fundamental frequency of the support falls below these critical values, implying that most supports will amplify the ground acceleration by more than twice its value.

We introduced the concept of "spectral acceleration amplification" to elucidate the relationship between dynamic amplification and inherent dynamic features. The curves in Figure 12 depict spectral acceleration amplifications for various support types and housing, showcasing peaks at frequencies corresponding to multiple higher-order vibration modes within the 2–10 Hz frequency range.

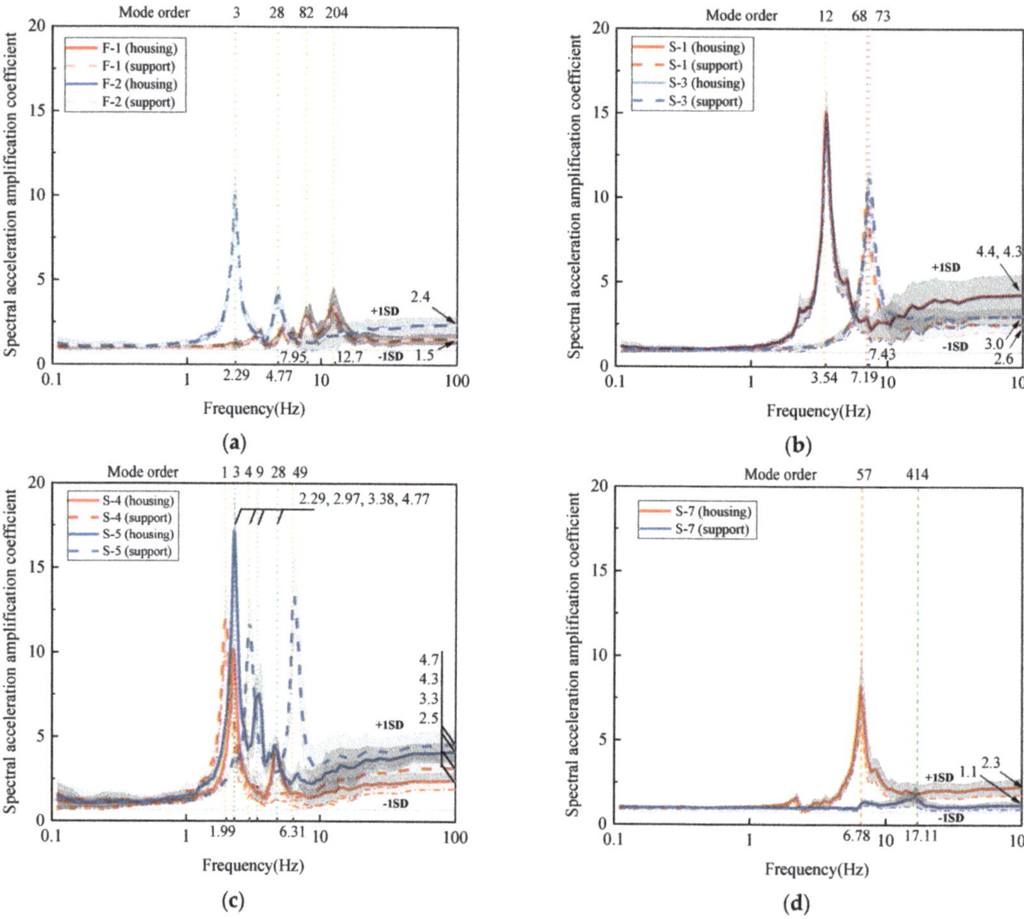

Figure 12. Spectral acceleration amplification curves of supports and housing. (**a**) F-1 and F-2; (**b**) S-1 and S-3; (**c**) S-4 and S-5; (**d**) S-7.

Key observations from Figure 12 include:

(1) The spectral acceleration amplification curve at high frequencies (100 Hz) represents the average acceleration amplification coefficient in the time domain.

(2) Prominent peaks at frequencies corresponding to multiple higher-order vibration modes indicate the influence of these modes on the dynamic amplification effect of supports and housing within the 2–10 Hz frequency range.

3.2. Three Retrofit Methods and Comparison of Results

Through the in-depth analysis of the dynamic amplification effect of the support in Section 3.1, it becomes evident that the dynamic amplification effect of the support on the pipeline cannot be ignored, with the acceleration amplification coefficient generally surpassing 2.0. Moreover, while notable disparities exist in the acceleration amplification coefficients among various types of supports and housing at distinct positions, there is a fundamental inverse relationship between the dynamic amplification effect and the fundamental frequency of the supports. This correlation suggests that enhancing the fundamental frequency of the supports can effectively diminish the dynamic amplification coefficient. Consequently, the focus of this section is to regulate the support's dynamic amplification effect by adjusting its fundamental frequency or altering the connection between the support and the pipeline, thereby accomplishing the seismic performance objective of mitigating the GIL's critical responses. Three retrofit methodologies were assessed and compared.

The first two methods increase the frequency of the support by changing the material of the support or increasing the cross-sectional area of the support, respectively, to increase the fundamental frequency of the support to $\sqrt{2}$ and $\sqrt{5}$ times the original fundamental frequency, recorded separately as $\sqrt{2}f_{support}$ and $\sqrt{5}f_{support}$. The third approach entails converting all supports to fixed supports. The peak ground acceleration in the X-direction is taken as 0.62 g. Table 3 presents the acceleration amplification coefficients of various support tops and housings under different retrofit methods. It is apparent that elevating the fundamental frequency of the support effectively reduces the acceleration amplification coefficient at the top of the support, and a similar trend is observed for the acceleration amplification coefficient at the top housing of the support. Notably, after all supports are fixed, the acceleration amplification coefficients of the supports and housings exhibit a trend of convergence toward a median value compared to before the retrofit. For supports with an acceleration amplification coefficient exceeding 4.0, the coefficient decreases after the retrofit, whereas supports with a coefficient below 4.0 experience an increase in acceleration amplification after the retrofit. This trend can be attributed to the enhanced overall structural integrity resulting from the full fixation of the support-to-pipeline connection.

Table 3. Acceleration amplification coefficients of housing and support under different retrofit conditions.

Supports	Original Model		$\sqrt{2}f_{support}$		$\sqrt{5}f_{support}$		Fixed All Supports	
	Housing	Support	Housing	Support	Housing	Support	Housing	Support
F-1	1.53	1.53	1.24	1.22	1.12	1.09	1.60	1.62
F-2	2.43	2.41	2.42	2.40	1.97	1.95	2.75	2.73
S-1	4.39	2.57	3.78	2.00	3.74	1.34	4.23	4.41
S-3	4.32	3.05	3.72	2.18	3.67	1.36	4.29	4.29
S-4	2.47	3.26	2.51	3.55	2.03	2.85	2.80	2.81
S-5	4.25	4.68	4.00	4.27	3.31	2.81	3.62	3.64
S-7	2.26	1.10	2.25	1.08	2.11	1.08	1.18	1.18

Table 4 highlights the key seismic response peaks of the GIL, where "S" represents peak housing stress and "D" represents relative displacement of internal conductor joints. The findings reveal that the first two retrofit methods not only fail to diminish the peak stress response of the GIL housing by increasing the fundamental frequency of the support, but also result in varying degrees of increased peak housing stress. Conversely, the third retrofit method, entailing the conversion of all supports to fixed ones, effectively reduces the peak housing stress, with the average peak stress response remaining below the material limit stress. All three retrofit strategies contribute to the reduction in relative displacement at internal conductor corner joints, with the "fixing all supports" method yielding the most significant reduction in relative displacement.

The findings presented in Tables 3 and 4 suggest that while enhancing the fundamental frequency of the support reduces the dynamic amplification effect, it does not significantly improve the seismic performance of the GIL. On the other hand, altering the connection between all supports and pipelines to fixed connections may not entirely regulate the dynamic amplification effect, but it notably enhances the overall seismic performance of the GIL. These results imply that the dynamic amplification effect of the support might not be the decisive factor influencing the seismic performance of the GIL, necessitating further analysis to elucidate this outcome.

Table 4. Critical peak responses of GIL under different retrofit conditions.

Ground Motions	Original Model		$\sqrt{2}f_{support}$		$\sqrt{5}f_{support}$		Fixed All Supports	
	S (MPa)	D (mm)	S (MPa)	D (mm)	S (MPa)	D (mm)	S (MPa)	D (mm)
El Centro	206.3	30.70	239.91	42.82	186.14	42.22	170.48	23.57
RSN66	171.2	85.70	318.37	46.21	252.64	46.29	230.49	35.00
RSN82	236.45	76.51	318.49	39.99	358.64	38.11	202.45	26.16
RSN1527	212.98	78.13	272.84	60.33	202.88	59.63	216.87	40.44
Chichi	290	42.20	269.47	42.96	207.14	38.97	178.69	21.01
Arti-1	226.48	41.37	226.59	41.04	226.63	40.03	173.43	23.84
Arti-2	219.04	42.88	252.43	41.87	219.56	42.87	206.79	22.99
Average	223.21	56.78	271.16	45.03	236.23	44.02	197.03	27.57

3.3. *Key Factors Influencing GIL Peak Responses*

This section primarily investigates the underlying reasons for the differing impacts of various retrofit methods on the peak stress response of the GIL. Given that the most significant peak stress responses of the housing and maximum relative displacement of the internal conductor joint are located at the pipeline corner between supports S-4 and S-5, special attention is warranted for the relative displacement or acceleration responses between these two positions. Figure 13a shows the scatter plot of the peak relative displacements of pipelines at S-4 and S-5 and the peak stress of the GIL housing under seven sets of seismic ground motions, including the original model and three retrofit methods. The circular markers denote the average values under these seismic conditions. Evidently, the above results indicate that the relative displacement response between S-4 and S-5 is a key factor affecting the peak response of the GIL housing. To provide a comprehensive understanding of how different retrofit measures influence relative displacement responses between S-4 and S-5 and subsequently reshape the overall seismic performance of the GIL, Figures 13b and 14, respectively, depict the power spectrum of relative displacement and the displacement cloud map of corresponding vibration modes.

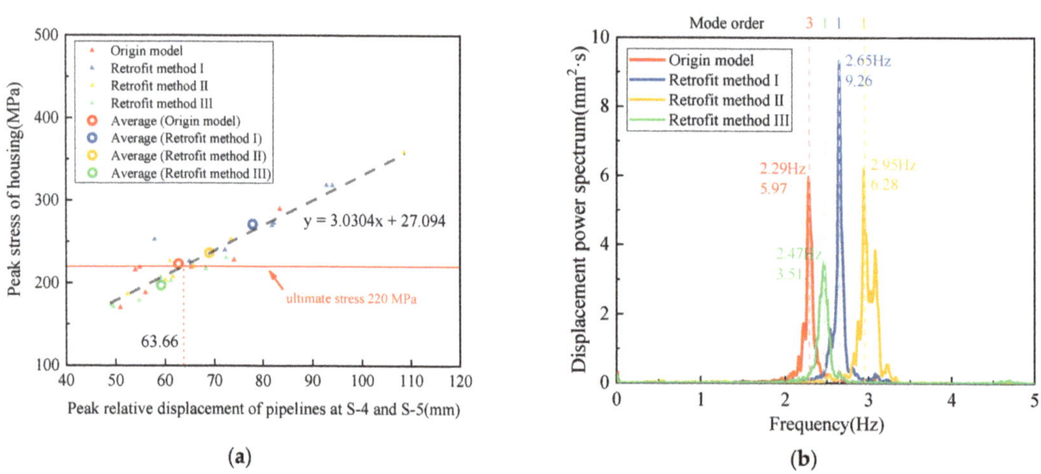

Figure 13. Key factors influencing GIL peak responses. (**a**) Relationship of peak stress of housing and relative displacement between S-4 and S-5; (**b**) power spectrum of relative displacement between S-4 and S-5.

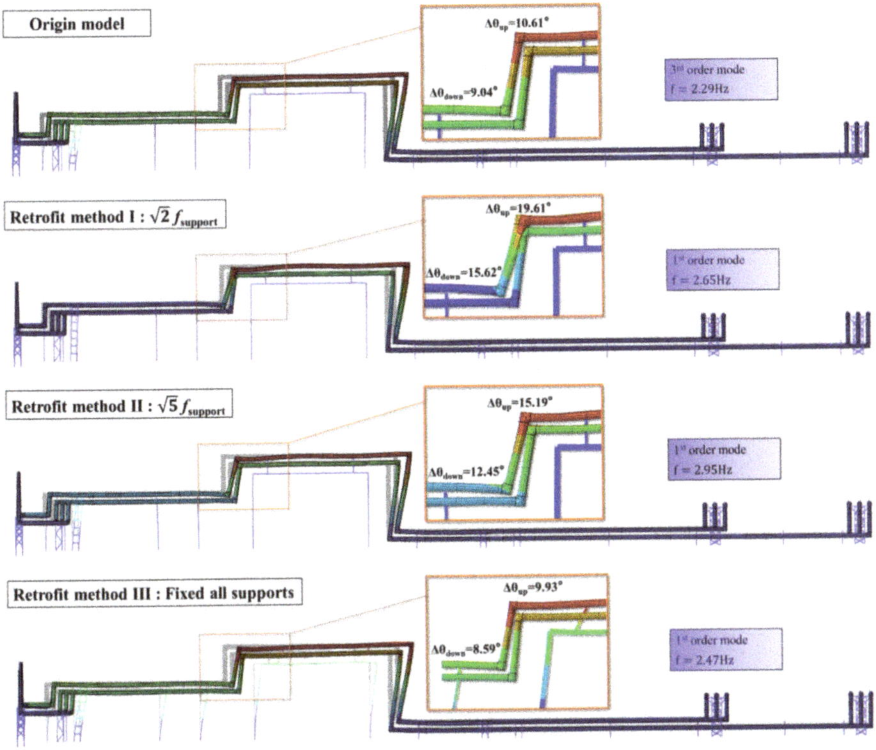

Figure 14. Comparison of target vibration modes under different retrofit conditions.

Figure 13b reveals a distinct pattern in the relative displacement power spectra of the original structure and the two other retrofit methods, showcasing a singular peak characteristic, except for the second retrofit method. The corresponding vibration mode order for each peak value is visibly denoted in Figure 14. Notably, the peak values of the relative displacement power spectrum for the three reinforced models are all aligned with the first-order vibration mode, while the peak values for the original structure correspond to the third-order vibration mode. These vibration modes are all associated with the X-direction translational motion of the pipeline, as illustrated in Figure 14. Furthermore, across all four cases, the relative magnitudes of the relative displacement power peaks mirror the relative magnitudes of the relative displacement peaks of S-4 and S-5, in the following order: retrofit method I > retrofit method II > original structure > retrofit method III. These insights emphasize the pivotal role played by the first-order vibration of pipelines in shaping relative displacement responses.

Figure 14 compares crucial mode shapes governing the relative displacement of supports S-4 and S-5, controlled across four distinct scenarios. Evidently, all vibration modes correspond to the first-order X-direction translational vibration of the pipeline, and the most significant deformation occurs at the corner of the pipe joint between supports S-4 and S-5. In the case of the third retrofit method, wherein all supports are fixed with the housing, supports S-4 and S-5 deform synchronously with the pipeline. Conversely, in the other three cases, sliding supports S-4 and S-5 remain undeformed. Noteworthy angular changes at the corners of key pipe joints are highlighted in the partially enlarged image. Similar to the peak values of the displacement power spectrum in Figure 13b, the relative magnitudes of the rotation angles in the key modes of different models in Figure 14 correspond to the relative magnitudes of the peak responses of the GIL for each model, which is also retrofit method I > retrofit method II > original model > retrofit method III.

Based on the analysis of the power spectrum of the relative displacement between supports S-4 and S-5 and the associated modal shape corresponding to the power spectrum peak, the following conclusions emerge:

1. Mitigating the relative displacement peak between supports S-4 and S-5 effectively diminishes the critical peak response of the GIL and enhances its seismic performance.
2. The peak relative displacement between supports S-4 and S-5 is governed by the first-order X-direction translational vibration mode of the pipeline, with the angle change value at the pipeline corner in this vibration mode displaying a positive correlation with the peak relative displacement between supports S-4 and S-5.
3. Insights from the deformation indicated that the pivotal factor affecting the seismic performance of the GIL is the deformation of the pipeline corner in the first mode of vibration, rather than the frequency corresponding to the first mode.

4. Seismic Vulnerability of GIL

Seismic vulnerability analysis can offer reasonably accurate failure probabilities for GILs under varying PGAs, thereby establishing a foundation for the seismic design and performance assessment of GILs [26,27]. The findings in Section 3 demonstrate that only retrofit method III can effectively diminish the critical seismic response of the GIL. Consequently, this section exclusively conducts seismic vulnerability analysis on the original structure of the GIL and retrofit method III.

4.1. Seismic Failure Criteria of GIL

Previous research has indicated that the seismic vulnerability of power equipment typically adheres to a logarithmic normal distribution [28–30]. When the peak ground acceleration (PGA) is x, the failure probability of a GIL can be represented by Formula (1).

$$P(PGA = x) = \Phi\left(\frac{(\ln x - \theta)}{\beta}\right) \tag{1}$$

where Φ denotes the cumulative distribution function of the standard normal distribution. θ and β represent the mean and variance, respectively. Based on the fitting results, the seismic vulnerability curve of the GIL can be derived. To obtain the fitting results, it is crucial to acquire a specific number of sample points, namely the failure probabilities of the GIL under varying PGAs. Initially, this study selected 30 ground motion records, with the screening conditions and average acceleration response spectra of these seismic records outlined in the following text. Subsequently, these 30 seismic records were normalized with amplitudes of 0.2 g, 0.4 g, 0.6 g, and 0.8 g, allowing for the calculation of the GILs' failure probability under these four sets of PGAs. The determination of the failure probability necessitates a clear understanding of the failure index.

Based on the previous research findings, two distinct failure modes of a GIL during seismic events have been identified: strength failure, indicated by a peak stress response exceeding the ultimate stress, and displacement exceeding the limit, leading to failure due to excessive relative displacement at the joint of the internal conductor corner. For the strength failure mode, the stress safety factor concept, as outlined in specification [20], is introduced. This implies that the ratio of ultimate stress to the actual stress response peak should not be less than 1.67, with an associated failure index. Consequently, the following three failure indicators have been established for GILs.

① The peak stress of the housing exceeds 60% of the limit stress (the factor of safety is 1.67, that is, 131.74 MPa), which is considered to cause damage.

② The peak stress of the housing exceeds that of the limit stress, 220 MPa, which is considered to cause damage.

③ The maximum displacement of the inner conductor joint exceeds 48 mm (limit value), which is considered to cause damage.

Among them, failure criteria ① and ② are considered from the perspective of strength and therefore become strength failure criteria, while failure criterion ③ is called the displacement failure criterion.

4.2. Selected Ground Motion Records

In accordance with the Chinese regulations, [23], the GIL site falls under the Class II category, with a shear wave velocity range of 140~250 m/s at a 30 m depth [31–33] and a characteristic period of 0.45 s. Utilizing these parameters, 30 natural seismic records were curated from the PEER database. Figure 15 illustrates their acceleration response spectra, average response spectra, and specification demand spectra, assuming a damping ratio of 2%. Notably, the average acceleration response spectrum of the 30 seismic motions effectively encapsulates the demand spectrum within the frequency range associated with the primary vibration modes of the GIL.

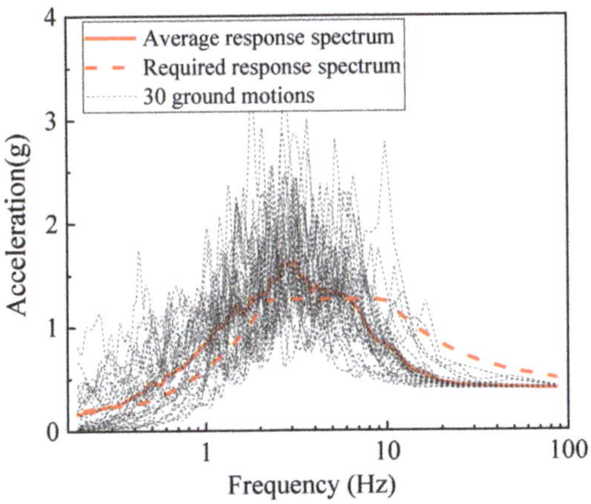

Figure 15. Average acceleration response spectrum of 30 sets of ground motion records and required response spectrum.

4.3. Seismic Fragility Curves of GIL

Under different failure criteria, the parameters in Formula (1) are determined through fitting based on the calculation results, as depicted in Table 5. Notably, for failure criterion ②, the failure probability of the reinforced GIL is 0; hence, parameter fitting was unnecessary. It is evident from the data that under each failure criterion, the average value for the reinforced GIL has noticeably increased compared to its unreinforced state. This increase signifies a significant decrease in the failure probability of the structure when the PGA is small.

Table 5. Fitting results of seismic vulnerability parameters.

Failure Criteria	Origin Model of GIL		Reinforced GIL	
	θ	β	θ	β
①	−0.84	0.23	0.02	0.94
②	−0.23	0.33	/	/
③	−0.61	0.36	−0.12	0.46

For the GIL and reinforced GIL models, the seismic vulnerability curves fitted based on the calculation results are shown in Figure 16. The triangular markers on the vulnerability curve signify the failure probability, calculated by dividing the count of seismic waves causing failure by the total waves across time history calculations at PGAs of 0.2, 0.4, 0.6, and 0.8 g.

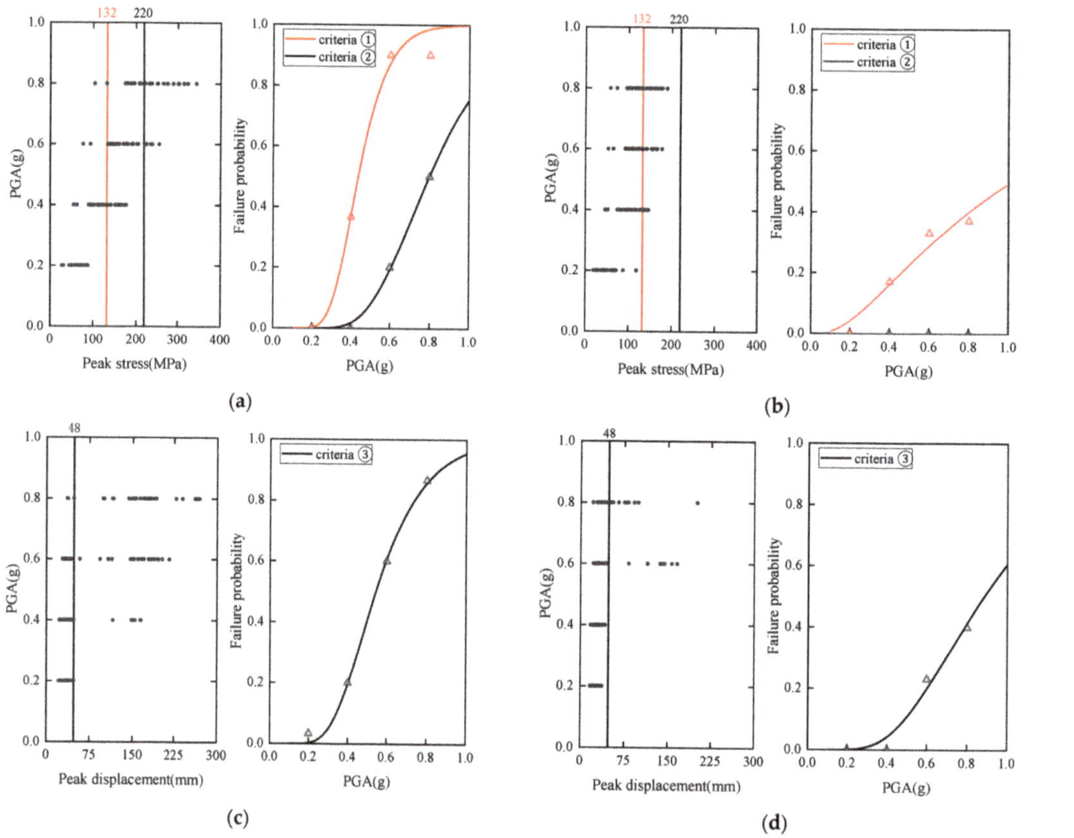

Figure 16. Seismic vulnerability analysis results of GIL before and after retrofit. (**a**) GIL under strength failure criterion; (**b**) reinforced GIL under strength failure criterion; (**c**) GIL under displacement failure criterion; (**d**) reinforced GIL under displacement failure criterion.

Figure 17 shows the seismic vulnerability curves of the GIL system before and after retrofitting under different failure criteria.

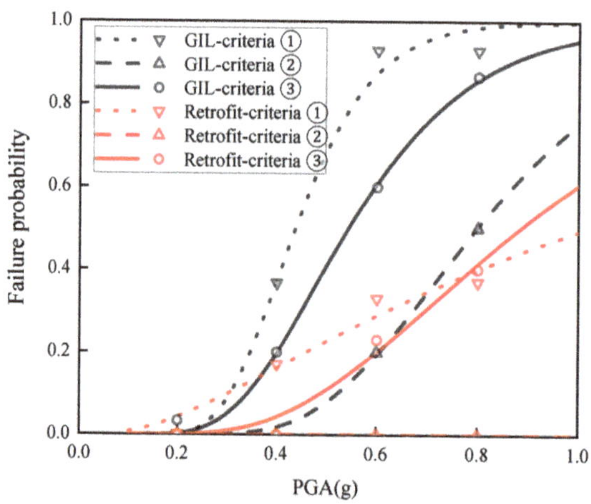

Figure 17. Seismic vulnerability curves of GIL before and after retrofit under different failure criteria.

The following conclusions can be drawn from Figures 16 and 17:

1. It can be clearly seen that under any failure criterion, the seismic vulnerability of the reinforced GIL is significantly reduced. Under failure criterion ①, when the PGA is 0.4 g, the failure probabilities of the GIL before and after the retrofit are 0.37 and 0.17, respectively, indicating a 53.7% reduction. When the PGA is 0.6 g, the failure probabilities of the GIL before and after the retrofit are 0.93 and 0.33, respectively, indicating a 64.5% reduction. Under failure criterion ②, when the PGA is 0.4 g, the failure probability of the GIL before and after the retrofit is 0. When the PGA is 0.6 g, the failure probabilities of the GIL before and after the retrofit are 0.2 and 0, respectively, indicating a 100% reduction. Under failure criterion ③, when the PGA is 0.4 g, the failure probabilities of the GIL before and after the retrofit are 0.2 and 0, respectively, indicating a 100% reduction. When the PGA is 0.6 g, the failure probabilities of the GIL before and after the retrofit are 0.6 and 0.23, respectively, indicating a 61.7% reduction.
2. Before the retrofit, the seismic vulnerability of the GIL under the strength failure criterion is higher than that under the displacement failure criterion. After the retrofit, the GIL seismic vulnerability curves corresponding to the two failure criteria intersect at a PGA of 0.8 g. When the PGA is less than 0.8 g, the strength failure criterion dominates, and when the PGA is greater than 0.8 g, the displacement failure criterion dominates.

5. Discussion

Currently, gas-insulated transmission lines (GIL) are increasingly prevalent in ultra-high-voltage converter power plants. However, with the constant threat of earthquakes, there is a growing need to enhance the seismic performance evaluation and retrofitting of GILs to ensure the safety of critical power transmission and transformation nodes during seismic events. This article conducts a comprehensive study on GILs, focusing on three key aspects: seismic performance, seismic retrofit methods, and seismic vulnerability, leading to the following conclusions:

(1) The research begins by developing a detailed finite element model for the GIL, accounting for its internal structure and simulating the connections between the TPI, the housing, and the supports. The results indicate that the seismic weak point of the GIL is located at the corner where the axis height changes, and there are two types of failure modes, namely housing strength failure and displacement-exceeding-limit failure at the internal conductor joint.

(2) An essential finding is the significant dynamic amplification effect of the supports, with an acceleration amplification coefficient generally exceeding 2.0. Importantly, the study reveals that the larger the fundamental frequency, the smaller the acceleration amplification coefficient of the supports. To reinforce the GIL, two approaches are explored: increasing the fundamental frequency of the support and altering the connection method between the support and the housing. The results indicate that both retrofit methods can effectively reduce peak displacements at the internal conductor joint. While increasing the fundamental frequency of the support reduces the acceleration amplification coefficient, it paradoxically increases the peak stress response of the housing. Conversely, fixing all mobile supports to the housing has a limited impact on the acceleration amplification coefficient but significantly reduces the peak stress response of the housing. These findings suggest that the dynamic amplification effect of the support is not the sole decisive factor influencing a GIL's seismic performance.

(3) The study delves further into the relationship between the relative displacement peak between key positions (S-4 and S-5) and the peak stress response of the GIL housing. It reveals a linear correlation, indicating that the peak housing stress increases proportionally with the rise in the relative displacement peak between these positions. The relative displacement between S-4 and S-5 primarily follows the first-order translational vibration mode of the pipeline in the X-direction, irrespective of the frequency

associated with this vibration mode. Instead, it is solely dictated by the deformation of the pipeline angle between S-4 and S-5 within this vibration mode.

(4) The study reinforces a GIL by fixing the connections between the sliding supports and the housing. The seismic vulnerability of the GIL is substantially reduced after the retrofit, particularly when the PGA surpasses 0.4 g. The seismic vulnerability of the GIL after the retrofit is found to decrease by over 50% compared to its pre-retrofit state, underscoring the effectiveness of this retrofit approach in minimizing the system's susceptibility to seismic events.

In conclusion, the research establishes a foundational understanding of the seismic performance of GILs in ultra-high-voltage converter power substations and similar overhead pipelines in earthquake-prone areas. It introduces innovative reinforcement measures and uncovers that the relative displacement between key sections, rather than the support acceleration amplification effect, is the pivotal factor influencing the seismic performance of GILs. The limitations in this study include the exclusive consideration of one GIL layout, neglecting the impacts of height difference and diverse support arrangements. Additionally, vulnerability studies did not encompass intensity measure (IM) indicators beyond PGA, and earthquake-induced electrical function failure modes were not addressed.

Author Contributions: Writing—original draft, X.L.; Writing—review & editing, Q.X. and J.W. All authors have read and agreed to the published version of the manuscript.

Funding: This research was funded by National Key R&D Program of China grant Number: 2018YFC0809400 and The APC was funded by Natural Science Foundation of Jiangsu Province grant Number: BK20230400.

Data Availability Statement: Data are contained within the article.

Conflicts of Interest: The authors declare no conflict of interest.

References

1. Zebouchi, N.; Haddad, M.A. A Review on Real-Size Epoxy Cast Resin Insulators for Compact High Voltage Direct Current Gas Insulated Switchgears (GIS) and Gas Insulated Transmission Lines (GIL)—Current Achievements and Envisaged Research and Development. *Energies* **2020**, *13*, 6416. [CrossRef]
2. Koch, H.; Schuette, A. Gas insulated transmission lines for high power transmission over long distances. *Electr. Power Syst. Res.* **1998**, *44*, 69–74. [CrossRef]
3. Khalil, M.; Ruggieri, S.; Tateo, V.; Nascimbene, R.; Uva, G. A numerical procedure to estimate seismic fragility of cylindrical ground-supported steel silos containing granular-like material. *Bull. Earthq. Eng.* **2023**, *21*, 5915–5947. [CrossRef]
4. Işık, E. Structural Failures of Adobe Buildings during the February 2023 Kahramanmaraş (Türkiye) Earthquakes. *Appl. Sci.* **2023**, *13*, 8937. [CrossRef]
5. Işık, E.; Avcil, F.; Büyüksaraç, A.; İzol, R.; Arslan, M.H.; Aksoylu, C.; Harirchian, E.; Eyisüren, O.; Arkan, E.; Güngür, M.Ş.; et al. Structural damages in masonry buildings in Adıyaman during the Kahramanmaraş (Turkiye) earthquakes (Mw 7.7 and Mw 7.6) on 06 February 2023. *Eng. Fail. Anal.* **2023**, *151*, 107405. [CrossRef]
6. Bülbül, M.A.; Harirchian, E.; Işık, M.F.; Aghakouchaki Hosseini, S.E.; Işık, E. A Hybrid ANN-GA Model for an Automated Rapid Vulnerability Assessment of Existing RC Buildings. *Appl. Sci.* **2022**, *12*, 5138. [CrossRef]
7. Karamanos, S.A.; Sarvanis, G.C.; Keil, B.D.; Card, R.J. Analysis and Design of Buried Steel Water Pipelines in Seismic Areas. *J. Pipeline Syst. Eng. Pract.* **2017**, *8*, 04017018. [CrossRef]
8. Tian, Y.; Filiatrault, A.; Mosqueda, G. Seismic Response of Pressurized Fire Sprinkler Piping Systems II: Numerical Study. *J. Earthq. Eng.* **2014**, *19*, 674–699. [CrossRef]
9. Tian, Y.; Filiatrault, A.; Mosqueda, G. Seismic Response of Pressurized Fire Sprinkler Piping Systems I: Experimental Study. *J. Earthq. Eng.* **2014**, *19*, 649–673. [CrossRef]
10. Hosseini, M.; Ghalyani, E.; Ghorbani Amirabad, N. Development of double-variable seismic fragility functions for oil refinery piping systems. *J. Loss Prev. Process Ind.* **2020**, *68*, 104259. [CrossRef]
11. Liu, A.; Chen, K.; Wu, J. State of art of seismic design and seismic hazard analysis for oil and gas pipeline system. *Earthq. Sci.* **2010**, *23*, 259–263. [CrossRef]
12. Tsinidis, G.; Di Sarno, L.; Sextos, A.; Furtner, P. A critical review on the vulnerability assessment of natural gas pipelines subjected to seismic wave propagation. Part 2: Pipe analysis aspects. *Tunn. Undergr. Space Technol.* **2019**, *92*, 103056. [CrossRef]
13. Tsinidis, G.; Di Sarno, L.; Sextos, A.; Furtner, P. A critical review on the vulnerability assessment of natural gas pipelines subjected to seismic wave propagation. Part 1: Fragility relations and implemented seismic intensity measures. *Tunn. Undergr. Space Technol.* **2019**, *86*, 279–296. [CrossRef]

14. Zhao, K.; Jiang, N.; Zhou, C.; Li, H.; Cai, Z.; Zhu, B. Dynamic behavior and failure of buried gas pipeline considering the pipe connection form subjected to blasting seismic waves. *Thin-Walled Struct.* **2022**, *170*, 108495. [CrossRef]
15. Jahangiri, V.; Shakib, H. Seismic risk assessment of buried steel gas pipelines under seismic wave propagation based on fragility analysis. *Bull. Earthq. Eng.* **2017**, *16*, 1571–1605. [CrossRef]
16. De Risi, R.; De Luca, F.; Kwon, O.S.; Sextos, A. Scenario-Based Seismic Risk Assessment for Buried Transmission Gas Pipelines at Regional Scale. *J. Pipeline Syst. Eng. Pract.* **2018**, *9*, 04018018. [CrossRef]
17. Rashidov, T.R.; Yuldashev, T.; Bekmirzaev, D.A. Seismodynamics of Underground Pipelines with Arbitrary Direction of Seismic Loading. *Soil Mech. Found. Eng.* **2018**, *55*, 243–248. [CrossRef]
18. Sultanov, K.S.; Vatin, N.I. Wave Theory of Seismic Resistance of Underground Pipelines. *Appl. Sci.* **2021**, *11*, 1797. [CrossRef]
19. Sun, D.; Guo, E.; Wu, H.; Liu, Z.; Mao, C.; Zhang, H.; Sun, X. Shaking Table Test of a Full-Scale RC Frame Structure with an Indoor Gas Piping System. *Adv. Civ. Eng.* **2022**, *2022*, 6869621. [CrossRef]
20. Blasi, G.; Perrone, D.; Aiello, M.A.; Pecce, M.R. Seismic performance assessment of piping systems in bare and infilled RC buildings. *Soil Dyn. Earthq. Eng.* **2021**, *149*, 106897. [CrossRef]
21. Kim, S.W.; Jeon, B.G.; Ahn, S.W.; Wi, S.W. Seismic behavior of riser pipes with pressure and groove joints using an in-plane cyclic loading test. *J. Build. Eng.* **2021**, *34*, 101911. [CrossRef]
22. *GB 50260-2013*; Ministry of Housing and Urban-Rural Development of the People's Republic of China (MOHURD), Code for Seismic Design of Electrical Installations. Ministry of Housing and Urban-Rural Development: Beijing, China, 2013.
23. *GB 50011-2016*; Ministry of Housing and Urban-Rural Development of the People's Republic of China (MOHURD), Code for Seismic Design of Buildings. Ministry of Housing and Urban-Rural Development: Beijing, China, 2016.
24. *GB 50191-2012*; Ministry of Housing and Urban-Rural Development of the People's Republic of China (MOHURD), Code for Seismic Design of Special Structures. Ministry of Housing and Urban-Rural Development: Beijing, China, 2012.
25. *GB 50981-2014*; Ministry of Housing and Urban-Rural Development of the People's Republic of China (MOHURD), Code for Seismic Design of Mechanical and Electrical Equipment. Ministry of Housing and Urban-Rural Development: Beijing, China, 2014.
26. Kwag, S.; Ryu, Y.; Ju, B.S. Efficient Seismic Fragility Analysis for Large-Scale Piping System Utilizing Bayesian Approach. *Appl. Sci.* **2020**, *10*, 1515. [CrossRef]
27. Ju, B.S.; Gupta, A.; Ryu, Y. Seismic Fragility of Steel Piping System Based on Pipe Size, Coupling Type, and Wall Thickness. *Int. J. Steel Struct.* **2018**, *18*, 1200–1209. [CrossRef]
28. Bender, J.; Farid, A. Seismic vulnerability of power transformer bushings. *Eng. Struct.* **2018**, *162*, 1–10. [CrossRef]
29. Wen, J.; Li, X.; Xie, Q. Cost-effectiveness of base isolation for large transformers in areas of high seismic intensity. *Struct. Infrastruct. Eng.* **2021**, *18*, 745–759. [CrossRef]
30. Baker, J.W. Efficient Analytical Fragility Function Fitting Using Dynamic Structural Analysis. *Earthq. Spectra* **2015**, *31*, 579–599. [CrossRef]
31. Dobry, R.; Borcherdt, R.D.; Crouse, C.B.; Idriss, I.M.; Joyner, W.B.; Martin, G.R.; Power, M.S.; Rinne, E.E.; Seed, R.B. New Site Coefficients and Site Classification System Used in Recent Building Seismic Code Provisions. *Earthq. Spectra* **2000**, *16*, 41–67. [CrossRef]
32. Borcherdt, R.D. Estimates of Site-Dependent Response Spectra for Design (Methodology and Justification). *Earthq. Spectra* **1994**, *10*, 617–653. [CrossRef]
33. Seed, H.B.; Ugas, C.; Lysmer, J. Site-dependent spectra for earthquake-resistant design. *Bull. Seismol. Soc. Am.* **1976**, *66*, 221–243. [CrossRef]

buildings

Article

Study of the Mechanical Properties of Beam-Column Joints in a New Type of Aluminum Alloy Portal Frame

Zhanqing Xing [1], Gang Wang [2,3,4,*], Xiaolin Lin [2], Jing Pang [1,*], Caiqi Zhao [3] and Qiaosheng Chen [4]

1 College of Science, Inner Mongolia University of Technology, Hohhot 010051, China; xingzhanqing@mecc.sinosteel.com
2 Guangzhou Yuehong Membrane Structure Engineering Co., Ltd., Guangzhou 511400, China; linxiaolin@gatents.com
3 School of Civil Engineering, Southeast University, Nanjing 211189, China; 101000815@seu.edu.cn
4 Shanghai Baoye Group Corp., Ltd., Shanghai 201900, China; bridge5188@163.com
* Correspondence: ww707509167@163.com (G.W.); pang_j@imut.edu.cn (J.P.)

Abstract: In the article, the semi-permanent aluminum alloy portal frame is used as the research background, beam-column joints are used as the research object, and experimental and numerical analyses are carried out. The influence of different bolt diameters and arch angles on the mechanical properties of beam-column joints under vertical load was analyzed using five sets of experiments. The experimental results show that the load–displacement curves of different bolt diameters in the elastic stage are basically consistent. After entering the plastic stage, the ultimate load first increases and then decreases, and the ultimate displacement is basically consistent. According to the experiment, there is no significant difference in the load–displacement curve when the arch angle increases from 90 degrees to 108 degrees. When the arch angle increases to 126 degrees, the stiffness and ultimate bearing capacity of the node under vertical load significantly increase. Then, a numerical analysis model was established to analyze the mechanical performance of beam-column joints under horizontal loads. The numerical analysis results indicate that under horizontal load, as the diameter of the bolt increases, the yield load, yield displacement, ultimate load, and ultimate displacement of the beam-column node exhibit no significant changes, and the change amplitude is minimal. When the beam-column node is subjected to horizontal loads, as the arch angle increases, the yield and ultimate displacement increase by 2.14 times and 2.78 times, respectively, and the yield and ultimate load decrease by 58% and 48%, respectively. Finally, a simplified design method for beam-column joints was proposed based on experiments and numerical analysis.

Keywords: aluminum alloy portal frame; beam-column joints; experimental study; numerical analysis; vertical load; horizontal load

Citation: Xing, Z.; Wang, G.; Lin, X.; Pang, J.; Zhao, C.; Chen, Q. Study of the Mechanical Properties of Beam-Column Joints in a New Type of Aluminum Alloy Portal Frame. *Buildings* **2023**, *13*, 2698. https://doi.org/10.3390/buildings13112698

Academic Editor: Hiroshi Tagawa

Received: 5 September 2023
Revised: 1 October 2023
Accepted: 3 October 2023
Published: 26 October 2023

1. Introduction

The commonly used types of joints in aluminum alloy spatial structures include gusset joints, bolt ball joints, cast aluminum joints, and hub joints, as shown in Figure 1. So far, some researchers have conducted a series of studies on these aluminum alloy joints and achieved excellent results.

Guo et al. [1–3] conducted an experimental study on the out-of-plane bearing capacity of plate joints and obtained the failure modes of the joints under different plate thicknesses. They found that as the plate thickness increased, the stiffness of the joints increased. They proposed a four-line model of semi-rigid out-of-plane plate joints. Then, through theoretical analysis, the bending stiffness and corresponding critical bending moment values of the joints in each stage of the four-line model were derived. Finally, a numerical model was established to simulate the stiffness of the bolt fixation stage and the hole wall pressure bearing stage, and based on the numerical analysis results, a formula for the out-of-plane bending stiffness of the joint was obtained. Then, based on out-of-plane research, a semi-rigid four-line model of in-plane aluminum alloy plate joints was obtained with numerical simulation. Guo et al. [4] conducted experimental research on the out-of-plane hysteresis performance of aluminum alloy plate joints. The experimental results showed that due to the gap between the bolt hole and the bolt rod, the hysteresis curve was not quite complete, and the torque relative rotation hysteresis curve showed that the joint had good energy dissipation capacity. Moreover, as the thickness of the joint plate increased, the command performance of the joint gradually improved. Ma et al. [5] proposed a new type of column plate joint based on the traditional aluminum alloy plate joint and obtained the bending moment angle curves of the column plate joint around the strong axis, weak axis, and torsion directions. Then, they were introduced into the grid shell beam element model. The analysis results showed that the column plate stage is still in the elastic stage when the grid shell is unstable. Chen et al. [6] conducted experimental research on the out-of-plane hysteresis performance of plate joints and obtained the hysteresis curve, ductility ratio, and energy dissipation rate of plate joints. In addition, they were compared with numerical simulation results and were found to be consistent. Liu et al. [7] conducted experimental research on the low cycle fatigue performance of plate joints, obtained the low cycle fatigue failure mechanism and fatigue life of plate joints, and then conducted numerical simulation. The numerical simulation results were significantly consistent with the experimental results. Finally, a local feature based low cycle fatigue life prediction method was proposed. Zhao et al. [8,9] improved the plate joint and proposed two new types of aluminum alloy joints. Through experimental research, it was found that both types of joints have good mechanical properties and can be used for single-layer aluminum alloy lattice shell structures.

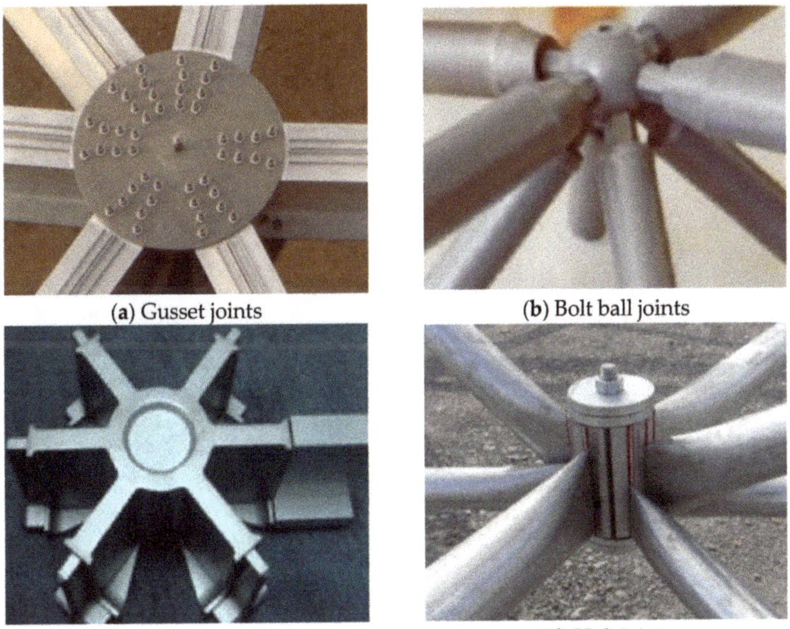
(**a**) Gusset joints (**b**) Bolt ball joints
(**c**) Cast aluminum joints (**d**) Hub joints

Figure 1. Diagram of aluminum alloy joints.

Hiyama et al. [10] proposed a formula for calculating the stiffness of aluminum alloy bolted ball joints through experimental and numerical simulation analysis. Liu et al. [11,12] studied a new type of aluminum alloy bolt spherical joint and evaluated it through tensile tests and numerical simulations. The tensile performance and failure mechanism of aluminum alloy bolt spherical joints was explored, and ultimately, a calculation formula for the tensile capacity of aluminum alloy bolt spherical joints was proposed. Shi et al. [13] conducted full-scale tests on three forms of cast aluminum joints in conjunction with the Chenshan Botanical Garden Greenhouse Exhibition Hall project. The test results showed that the cast aluminum joints had a greater flexural stiffness outside the plane and a smaller stiffness in the plane, making them typical semi-rigid joints. Their failure mode was brittle fracture failure. Subsequently, numerical simulations were conducted, and a simplified calculation formula for their bearing capacity was proposed. Cast aluminum joints can be widely used in aluminum alloy spatial structures. Sugizaki, K et al. [14,15] conducted experimental research on the basic mechanical properties of the cast aluminum joint, and the research results showed that adjusting the material properties of the component can ensure that the ultimate tensile bearing capacity of the joint is approximately the product of the tensile strength of the pipe and the cross-sectional area, which also indicates that this joint has an obvious semi-rigidity. Yonemaru et al. [16] conducted bending tests on aluminum alloy and carbon fiber-reinforced composite hub joint trusses, and the results showed that the overall truss buckling occurred due to the failure of the upper components.

The lightweight portal frame itself has the advantages of large building space, simple structural stress, and clear transmission path. The aluminum alloy portal frame structure combines many advantages brought by materials [17,18], such as lightweight, high strength, corrosion resistance, and high assembly rate. Therefore, aluminum alloy portal frame structures can be used not only in traditional permanent buildings such as workshops and warehouses, but also in temporary or semi-permanent buildings such as healthcare camps, exhibitions, sports events, and logistics warehousing. At present, there are relatively few research results on aluminum alloy portal frame joints. In this article, a new type of beam-column joint for aluminum alloy portal steel frames is proposed. Then, experimental research and numerical simulation analysis were conducted on this new type of beam-column joint. Finally, based on experiments and numerical simulations, a simplified design method for this type of joint was proposed.

2. Test Scheme

2.1. Specimens Design

To explore the mechanical properties of aluminum alloy portal frame beam-column joints, in this section, beam-column joint specimens are designed as shown in Figure 2. The beam-column joint specimens are made of H-shaped aluminum alloy members and double C-shaped double groove connectors tightly connected by bolts. Among them, the cross-sectional size of the aluminum alloy I-beam is H203 × 106 × 11 × 11 mm, and the channel steel connection is C181 × 47.5 × 5 × 10 mm. Six and four bolt holes are, respectively, set on the upper and lower flanges of the H-shaped aluminum alloy I-beam, and six bolt holes are set on the web of the I-beam. At the same time, bolt holes are set at the corresponding positions of the flange and web of the double groove connector. The diameter of the bolt holes is determined based on the corresponding bolt diameter (greater than 0.2 mm of the bolt diameter). The lengths of aluminum alloy I-beams are 1240 mm and 990 mm, respectively, and the lengths of channel steel connectors are 470 mm. The distribution spacing of bolts on the flange and web is 148 mm. The structural dimensions of the specimens are shown in Figure 2a. The aluminum alloy I-beams are cut and drilled in the factory, and the double groove connectors are welded from steel plates. The two are then transported to the laboratory for assembly and connection, as shown in Figure 2b.

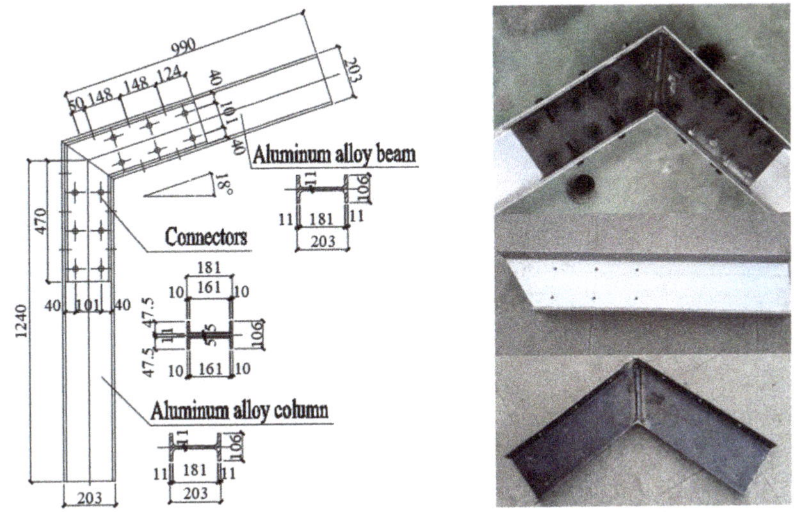

(**a**) Schematic diagram (**b**) Physical image

Figure 2. Test specimen.

For portal frames, the bolt diameter and arch angle of beam-column joints have a significant impact on the mechanical performance of their beam-column joints. Therefore, a total of 5 beam-column joint specimens were designed in this chapter, and the parameters of each specimen are detailed in Table 1. From this table, it can be seen that this experiment mainly designed beam-column joint specimens with different bolt diameters (8 mm, 14 mm, and 20 mm) and different arch angles (90°, 108°, and 126°) to explore the impact of different bolt diameters and arch angles on the mechanical properties of aluminum alloy portal frame beam-column joints. It should be noted that when studying the influence of bolt diameter or arch angle, other structural parameters of beam-column joints remain consistent.

Table 1. Detailed information of joints.

Specimen	Bolt Diameter (mm)	Arch Angle (°)	Beam Section Size (mm)	Connection Section Size (mm)
SJ-1	8	108	H203 × 106 × 11 × 11	2C181 × 47.5 × 5 × 10
SJ-2	14	108	H203 × 106 × 11 × 11	2C181 × 47.5 × 5 × 10
SJ-3	20	108	H203 × 106 × 11 × 11	2C181 × 47.5 × 5 × 10
SJ-4	20	90	H203 × 106 × 11 × 11	2C181 × 47.5 × 5 × 10
SJ-5	20	126	H203 × 106 × 11 × 11	2C181 × 47.5 × 5 × 10

2.2. *Testing System*

The loading site of this experiment is shown in Figure 3a. The experimental system mainly includes reaction frame (composed of reaction beam and column), jack, consolidation support, loading control instrument, and data acquisition instrument. During the experiment, the lower end of the specimen is first tightly connected to the consolidation support, and then, a vertical concentrated force is applied to the specimen through a jack, as shown in Figure 3b. Four bolts and end plates are used to securely connect the jack to the reaction beam, ensuring that the position of the loading point does not shift during the loading process, as shown in Figure 3c. The consolidation support is tightly connected to the foundation through vertically arranged threaded steel bars, while the bottom of the specimen is tightly connected to the consolidation support through three drainage flat bolts, ensuring the formation of consolidation constraint boundary conditions at the bottom of the specimen, as shown in Figure 3d. By using the above loading methods and support forms, the boundary conditions of aluminum alloy portal frame beam-column joints under vertical load can be simulated, thereby ensuring the effectiveness of the test results. During the experiment, the loading control instrument was used to control the loading speed. In this experiment, the loading speed was 1 mm/min, and the data collection instrument was used to record the test data at each moment.

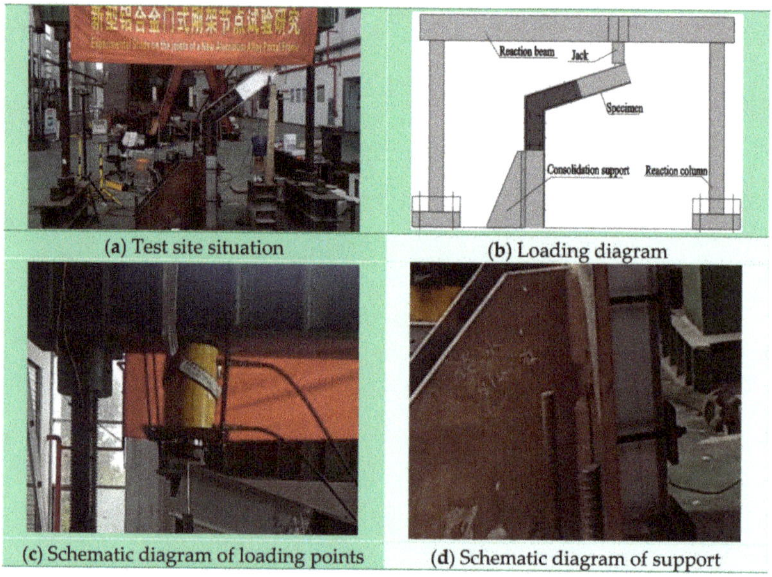

(**a**) Test site situation (**b**) Loading diagram

(**c**) Schematic diagram of loading points (**d**) Schematic diagram of support

Figure 3. Schematic diagram of test.

During the experiment, the strain and displacement data of the specimen at each moment are recorded using a data acquisition instrument, and the positions of each measuring point are shown in Figure 4. To obtain the strain variation pattern of H-type aluminum alloy beams and C-type channel steel connections under vertical load, a total of 10 sets of strain gauges were set at different parts of the cross-section, as shown in Figure 4a. Among them, strain gauges 1–5 are arranged on the cross-section of aluminum alloy beams, and gauges 6–10 are arranged on channel steel connectors. The No.1 strain gauge is arranged on the upper flange of the aluminum alloy beam, the 2nd, 3rd, and 4th are, respectively, arranged on the beam web from top to bottom, and the 5th is arranged on the lower flange of the aluminum alloy beam. Strain gauges 6 and 10 are, respectively, arranged on the upper and lower flanges on the left side of the channel steel connection, while gauges 7, 8, and 9 are evenly arranged on the web plate of the channel steel connection. The full cross-section strain changes in aluminum alloy beams and channel steel connections are monitored and recorded during the loading cycle using 10 strain gauges.

Two displacement meters are, respectively, arranged at the time loading point and the end of the channel steel connection, as shown in Figure 4b. The No. 1 displacement meter is set on the end side of the aluminum alloy beam flange to record the displacement changes at the end points of the aluminum alloy beam. The No. 2 displacement meter is set on the end side of the channel steel connector to record the displacement changes at the end of the channel steel connector.

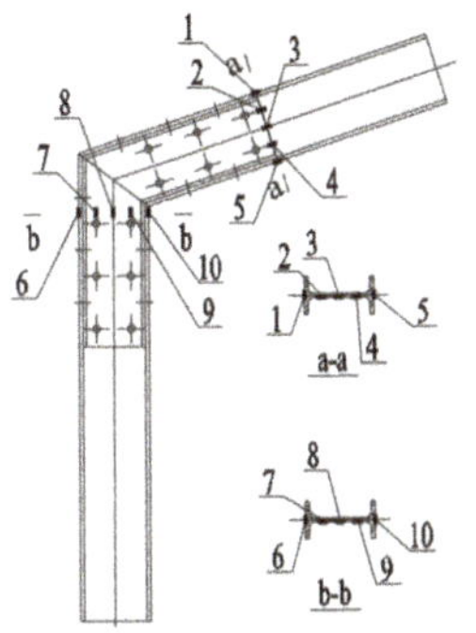

(a) Strain gauge layout plan

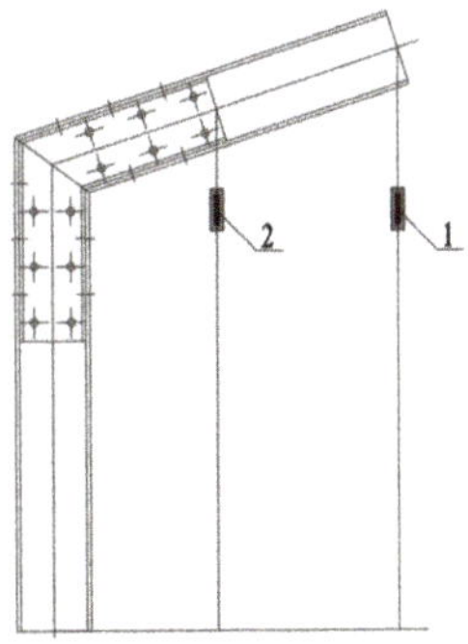

(b) Displacement meter layout plan

Figure 4. Schematic diagram of measurement point layout.

2.3. Material Property Test

The H-shaped I-beam of the test piece is made of 6061-T6 [19,20] aluminum alloy profile, and the double groove connector is made of Q235 steel. Samples from the same batch of aluminum and steel used in the test components are obtained and processed into standard specimens for material properties testing [21], as shown in Figure 5a. The main length of the material test piece is 250 mm, with both ends being 55 mm and 25 mm in length and width, and the middle section being 90 mm and 15 mm in length and width, respectively. The two ends and middle section are transitioned through a circular arc with a radius of 36 mm. The dimensions of the test piece are shown in Figure 5b. The material testing tensile machine tightly clamps the two ends of the material test piece, and then begins to apply axial tension, as shown in Figure 5c. The failure mode of the specimens is shown in Figure 5d, and all of them show tensile failure of the middle section. The results of the material properties test are summarized in Table 2.

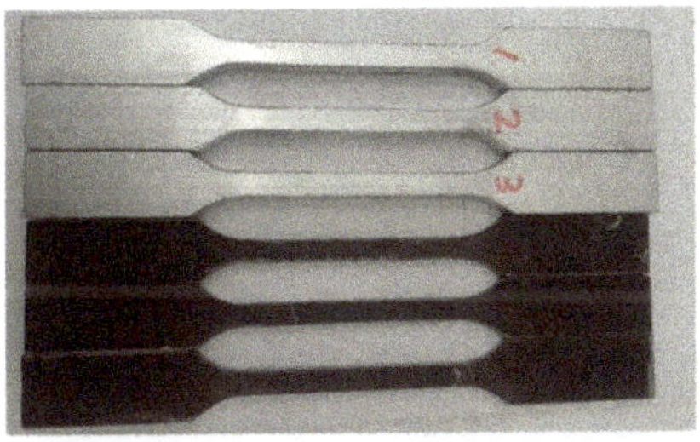

(a) Physical image of the specimen

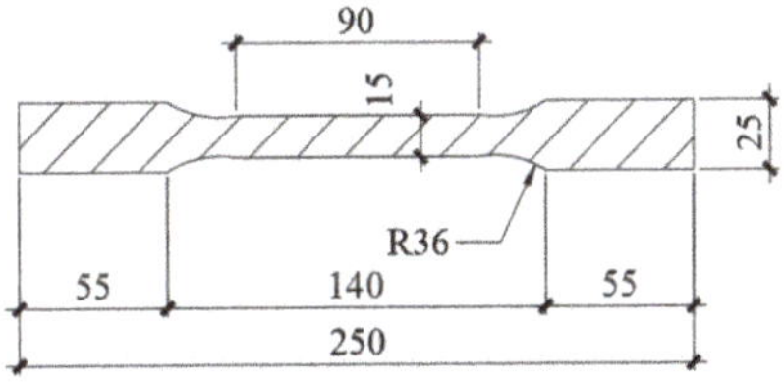

(b) Sample size diagram

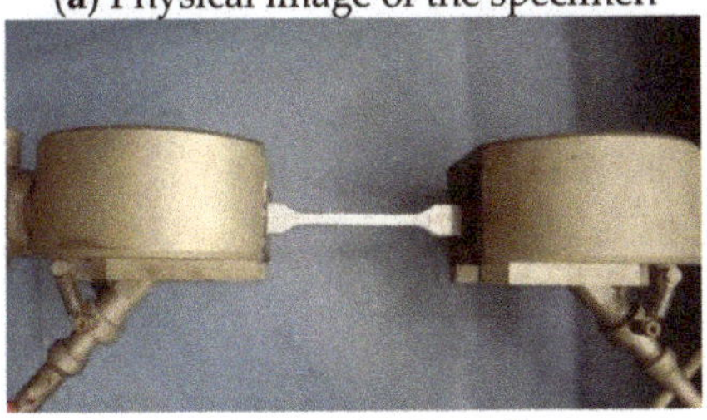

(c) Material test loading diagram

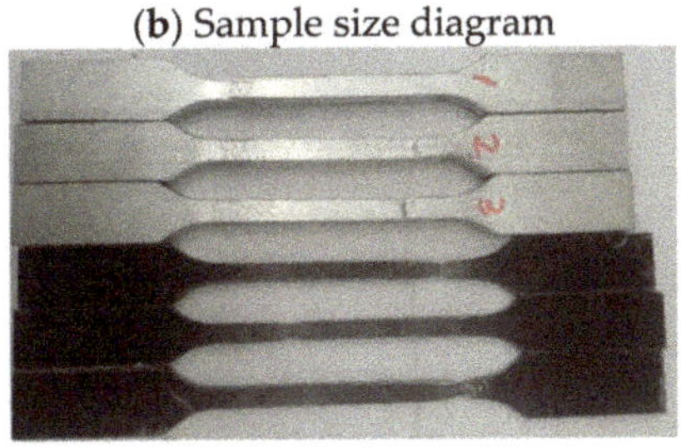

(d) Failure diagram of material specimens

Figure 5. Schematic diagram of material properties test.

Table 2. Material test results.

Material	Yield Strength MPa	Tensile Strength MPa	Elastic Modulus GPa
6061-T6	239	264	70.5
Q235	235	360	206
Bolt	887	992	204

3. Analysis of Test Results

3.1. Failure Characteristics

For beam-column joints with different bolt diameters, the failure state of the bolt hole wall needs to be focused on. Therefore, the bolts were removed after the specimen failed, and the state of the bolt hole wall is shown in Figure 6. When the bolt diameter is 8 mm, there is a significant extrusion deformation on the hole wall. When the bolt diameter is 14 mm, some bolt hole walls show slight compression deformation, while some hole walls show no significant compression. The bolt diameter increased to 20 mm, and the bolt hole walls remained intact without obvious signs of compression. From the above analysis, it can be seen that as the diameter of the bolt increases, the extrusion area of the bolt hole wall increases, and the extrusion deformation of the hole wall gradually decreases.

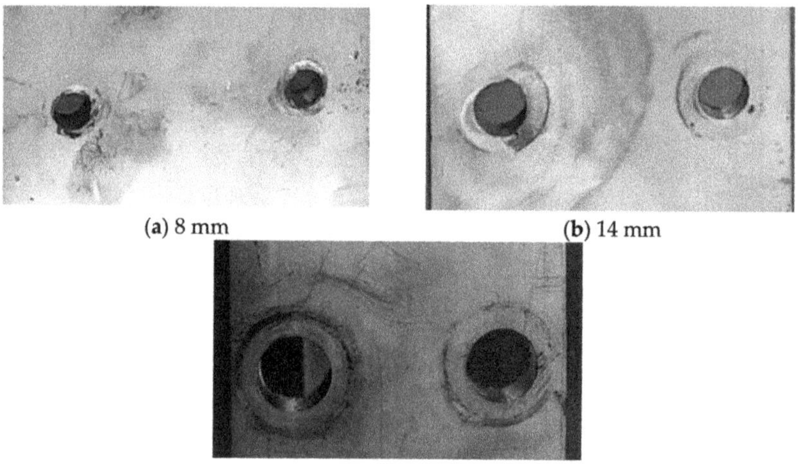

(**a**) 8 mm (**b**) 14 mm (**c**) 20 mm

Figure 6. Failure modes of different bolt diameters.

The failure modes of beam-column joint specimens with different arch angles are summarized in Figure 7. From this figure, it can be seen that the failure modes of the beam-column joint specimens are basically consistent at different arch angles. As the vertical load increases, the bending moment at the corner of the aluminum alloy portal frame beam-column joint gradually increases, causing buckling failure of the outer flange corner of the double groove connection, and is accompanied by an increase in the gap between the aluminum alloy beam flanges.

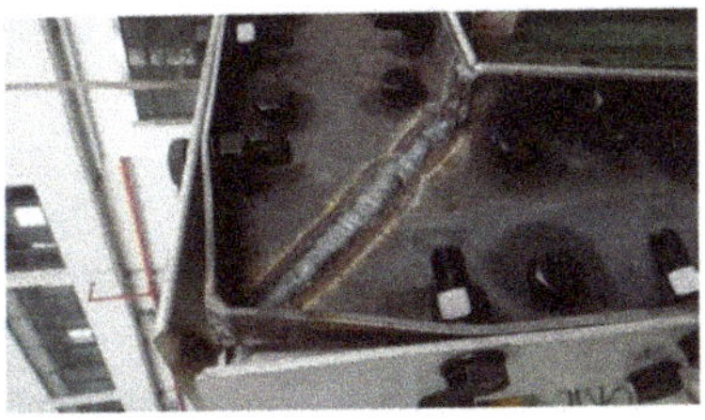

(**a**) 90°

(**b**) 108°

(**c**) 126°

Figure 7. Failure modes from different angles.

3.2. Strain Curve

In order to investigate the variation law of the full section strain of aluminum alloy beams and double groove connectors during vertical load loading, the loading cycle was divided into five specimen joints, and the strain distribution status of each joint section was extracted, as shown in Figure 8. Observing the strain graph curves at various time points, the observations made are as follows:

(1) The cross-sectional strain distribution of aluminum alloy beams and channel steel connectors of each specimen is basically consistent, which is due to the similar structural forms and consistent loading methods of each joint specimen;
(2) The strain distribution of aluminum alloy beams is similar to that of composite I-shaped sections under bending moment, showing a general pattern of larger strain on the upper and lower flanges and smaller strain on the web;
(3) The maximum strain of the channel steel connection occurs at the junction of the web and the upper and lower flanges, with the strain values of the upper and lower flanges in the middle and the strain in the middle of the web being the smallest.

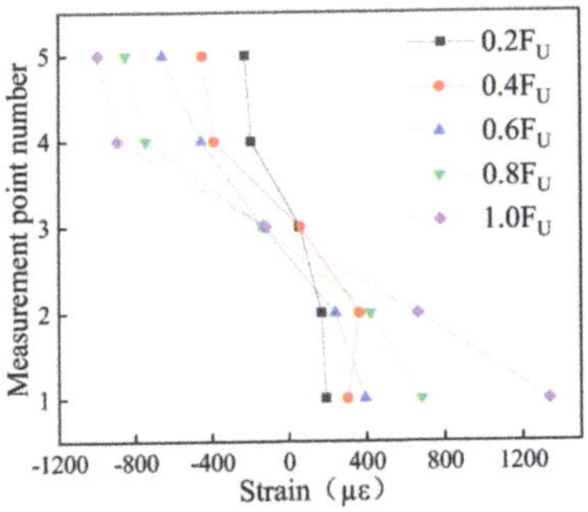

(**a**) Aluminum alloy beam of SJ-1

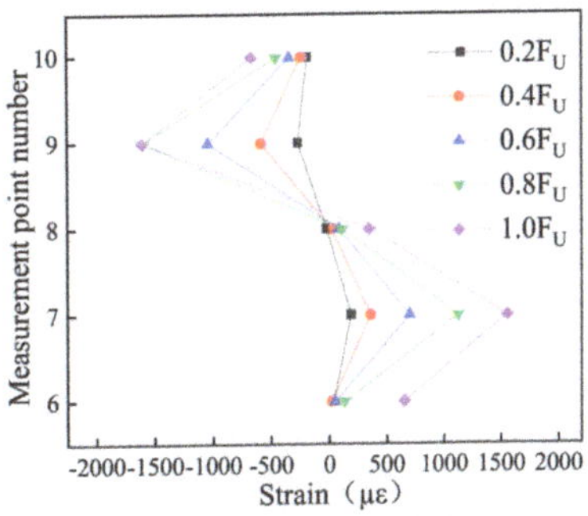

(**b**) Connector of SJ-1

Figure 8. *Cont.*

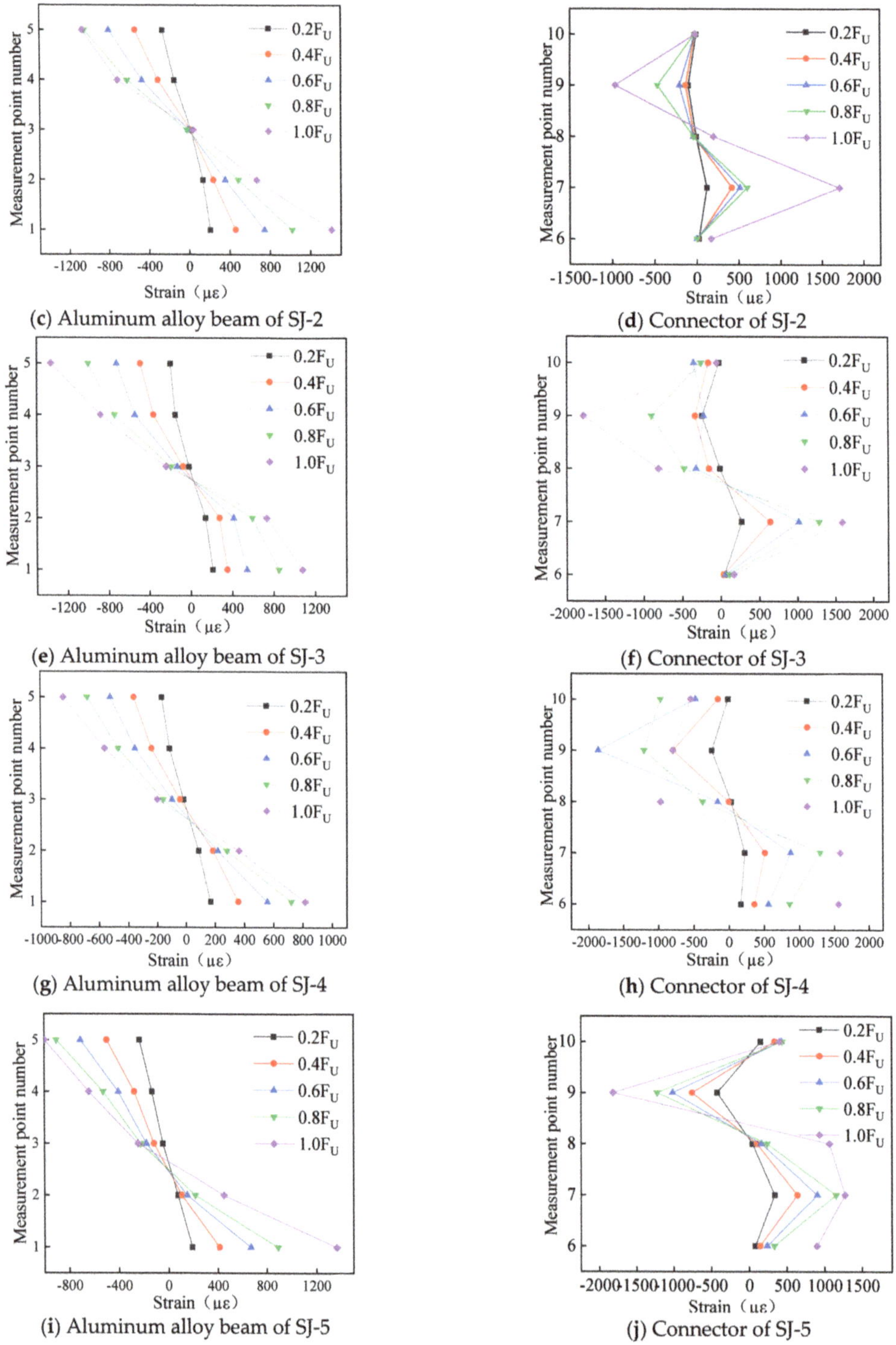

(c) Aluminum alloy beam of SJ-2
(d) Connector of SJ-2
(e) Aluminum alloy beam of SJ-3
(f) Connector of SJ-3
(g) Aluminum alloy beam of SJ-4
(h) Connector of SJ-4
(i) Aluminum alloy beam of SJ-5
(j) Connector of SJ-5

Figure 8. Schematic diagram of strain distribution.

The strain distribution of aluminum alloy beams basically indicates that the specimen mainly bears the action of bending moment under vertical load, and the position of strain 0 is not far below the center point of the section, indicating that the component simultaneously bears the action of shear force, but the shear force is relatively small. By analyzing the strain distribution pattern of double groove connectors, it can be concluded that the web of the double groove connector is responsible for transmitting most of the bending moment, while the flange transmits a small portion of the bending moment. Therefore, in joint design, the thickness of the web plate of the double groove connection should be appropriately increased.

3.3. Load–Displacement Curve

Based on the analysis factors of different bolt diameters and arch angles, the load−displacement curves of the specimens were summarized and compared, as shown in Figure 9. The load−displacement curve of aluminum alloy portal frame beam-column joints under vertical load mainly includes three stages: elastic stage, yield stage, and degradation stage. A detailed analysis of the load−displacement curves for different bolt diameters and arch angles based on three stages is conducted.

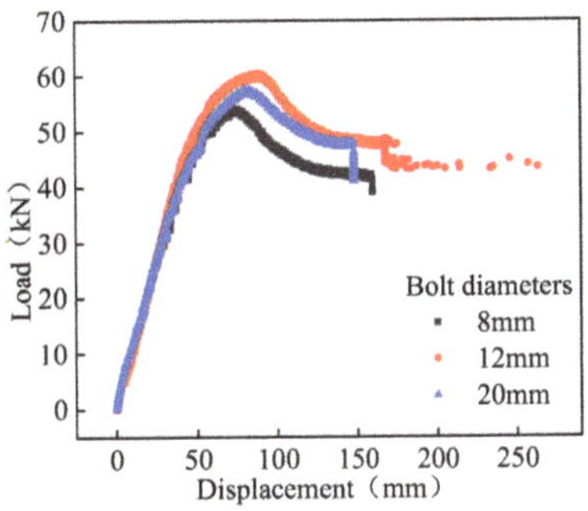

(**a**) Different bolt diameters

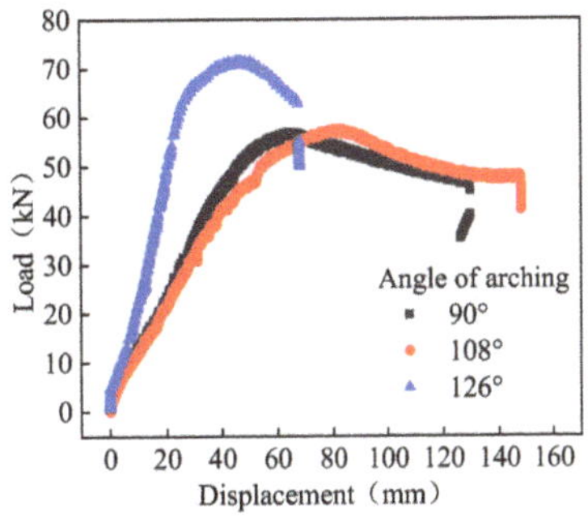

(**b**) Different arch angles

Figure 9. Load−displacement curve.

The load−displacement curves of different diameters are shown in Figure 9a. In the elastic stage, the load–displacement curves of different bolt diameters are basically consistent, and the yield load is close to 50 kN. After entering the plastic stage, the load–displacement curves of different diameters begin to show differences, meaning that the ultimate load first increases and then decreases, and the ultimate displacement is basically the same. Among them, the ultimate load is 52 kN for a diameter of 8 mm, 63 kN for a diameter of 12 mm, and 58 kN for a diameter of 20 mm. This is because as the diameter of the bolt increases, the shear-bearing capacity of the bolt group increases, resulting in an increase in the bearing capacity of the specimen. But, when the bearing capacity of the bolt group is greater than the bearing capacity of the net section of the rod, increasing the bolt diameter will not cause an increase in the bearing capacity of the specimen, but will instead cause a decrease in the net section area, leading to a downward trend in the bearing capacity of the specimen. In the degradation stage, the downward trend observed in load–displacement curves with different diameters is basically consistent, and the differences between the three curves are the same as those in the plastic stage, so we will not elaborate on them here.

The load−displacement curves for different arch angles are summarized in Figure 9b. When the arch angle increases from 90 degrees to 108 degrees, the load–displacement curve basically matches in the elastic stage, and there is a slight difference between the plastic stage and the degradation stage. When the arch angle increases to 126 degrees, there are significant differences in the load–displacement curve in the elastic stage, yield joint, and degradation stage, that is, the stiffness and ultimate bearing capacity of the specimen under vertical load significantly increase, with an increase of 30% in ultimate bearing capacity and a decrease of 35% in ultimate displacement. The reason for the above phenomenon is that when the arch angle increases to a certain extent, the force on the beam under vertical load changes from mainly bending to compression bending. Axial pressure can improve the bending performance of the beam to a certain extent, thereby reducing the vertical displacement caused by bending moment.

4. Establishment of Numerical Models

4.1. Numerical Model

Based on the geometric symmetry of aluminum alloy portal frame beam-column joints, a 1/2 joint model was established in ABAQUS with the cross-section of the center line of the aluminum alloy I-beam web as the symmetry plane, as shown in Figure 10a. The geometric dimensions of the numerical model are strictly consistent with the experimental components, in order to verify the effectiveness of the model compared with the experimental results. Each component adopts an eight-joint six-sided linear reduced integral element (C3D8R element). The grid size of aluminum alloy beams and connectors is 5 mm, the grid size of bolts is 2 mm, and the total number of grids in the node model is 65,000. During the modeling process, aluminum alloy beams, double groove connectors, and bolts will be divided into different component groups to facilitate model calculation and query of later calculation results.

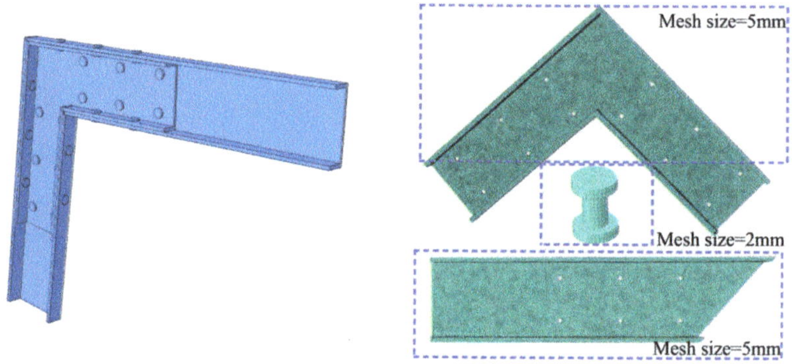

(a) Geometric model (b) Mesh situation

Figure 10. Joint numerical model.

The division of finite element mesh directly determines the accuracy and speed of calculation. The mesh division of this model is shown in Figure 10b. According to the structural and stress characteristics of the beam-column joints, the grid density should be appropriately increased at the corners and bolt connections of the model, and relatively small grid densities can be used for grid division at other positions, but it should be ensured that no less than two segments of grid are divided along the thickness direction.

When conducting numerical simulations, the selection of material models will directly determine the effectiveness of the calculation results. In the numerical model of beams and columns, there are mainly two types of materials, namely, aluminum alloy and steel. Based on the results of material properties tests, the Ramberg–Osgood model was used for aluminum alloy [22,23], and the double-line model was used for steel, as shown in Figure 11.

According to the loading scheme and constraint conditions of the test specimen, the boundary conditions of this model mainly include three aspects, namely, fixed end constraints, symmetric constraints, and contact settings [24]. The fixed end constraint is consistent with the experiment, that is, a consolidation constraint is set at the bottom of the column component of the beam-column joint specimen to constrain the displacement and rotation of the column bottom, as shown in Figure 12a. According to the geometric symmetry of the joint specimen, symmetric constraints are set on the central section of the web, as shown in Figure 12b. Relative sliding occurs between different components of the specimen, so it is necessary to set up a "limited slip" contact. The bolt rod and bolt hole wall only need to be set up with a normal "hard contact", while other contacts need to consider friction, that is, set a tangential "Coulomb friction" with a friction coefficient of 0.3. The coupling point is set at the center of the section at the loading end, which is used to apply the load, as shown in Figure 12c.

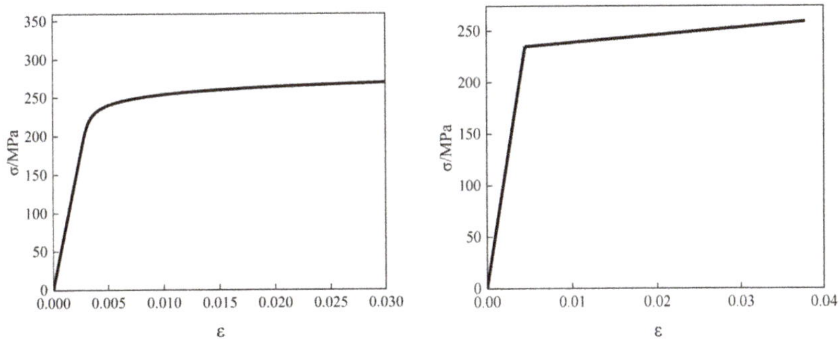

(**a**) Aluminum alloy (**b**) Steel

Figure 11. Material models.

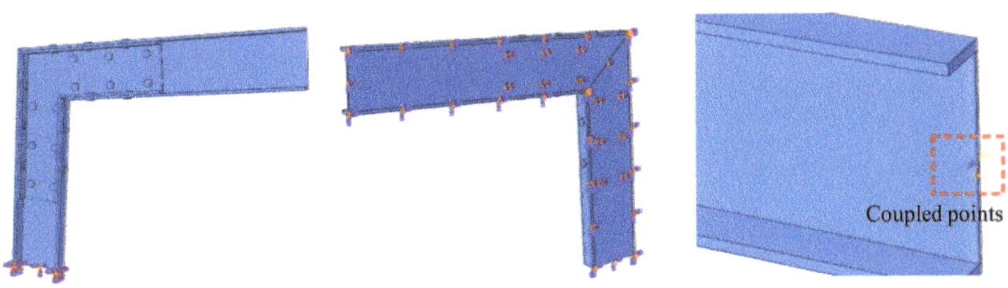

(**a**) Consolidation constraint (**b**) Symmetric constraint (**c**) Coupling point

Figure 12. Boundary conditions.

4.2. Model Validation

To verify the effectiveness of the numerical model, test specimens SJ-1 and SJ-5 were simulated, and the comparison of load–displacement curves is shown in Figure 13. From this figure, it can be seen that the load–displacement curves of the experiment and numerical simulation are relatively consistent, especially the variation law of the second type curve is basically consistent. The ultimate loads and displacements for experiments and numerical simulations are listed in Table 3. The comparison results show that the ultimate load error between the experiment and numerical simulation is only 4.77%, and the ultimate displacement error is only 4.01%. Obviously, this type of numerical model can be used for parameter analysis of aluminum alloy portal frame beam-column joints.

Table 3. Comparison of test and numerical simulation results.

Specimen	Type	Displacement (mm)	Load (kN)
SJ-1	Test	72.8	53.40
	FEA	72.8	58.17
SJ-5	Test	48.6	71.60
	FEA	48.6	67.59
Error		4.77%	4.01%

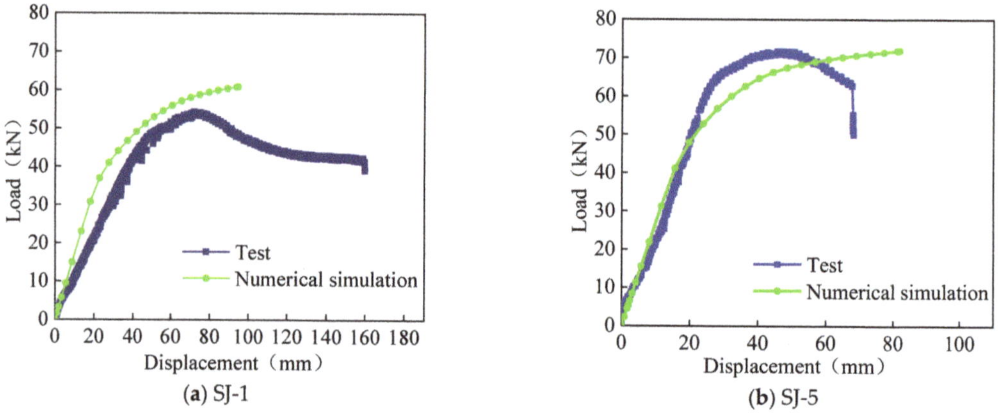

Figure 13. Comparison of load–displacement curves.

5. Parameterized Analysis of Numerical Simulation

5.1. Basic Model Parameters

Before conducting numerical simulation parameter analysis, the basic model information will be introduced, and subsequent models will adjust the response parameters based on the different analysis parameters of the basic model. The cross-sectional dimensions of the H-type aluminum alloy rod in the basic model are H200 × 100 × 8 × 10, the cross-sectional dimensions of the groove type connector are 2C180 × 46 × 5 × 10, the arch angle is 96 degrees, the bolt grade is 10.9, and the bolt diameter is 20 mm. The H-type aluminum alloy rod is made of 6061-T6 aluminum alloy (from China) profile, and the channel steel connector is made of Q235 steel (from China). On the basis of the basic model, a numerical analysis model is established for different bolt diameters, arch angles, and thickness of groove connectors, in order to analyze the influence of these factors on the mechanical properties of aluminum alloy portal frame beam-column joints. In the basic model, the load forms are divided into three types, namely, vertical concentrated force and horizontal concentrated force, which are used to analyze the force characteristics and deformation mechanism of this type of joint under the action of out-of-plane vertical load and in-plane horizontal load.

5.2. Mechanical Properties under Vertical Loads

The mechanical performance of aluminum alloy portal frame beam-column joints under vertical load was studied through experiments with different bolt diameters and arch angles. Here, numerical simulation was used to supplement and analyze the influence of arch angle and channel steel wall thickness on the vertical performance of this type of joint. The vertical load–displacement curves for different arch angles and channel steel wall thicknesses are summarized in Figure 14 and Tables 4 and 5:

(1) As the arch angle increases (from 96 degrees to 136 degrees), there is no obvious change in yield displacement and ultimate displacement, and the change amplitude is small. The yield load and ultimate load gradually increase (2.04 times and 1.90 times, respectively), and the failure characteristics gradually change from groove type connection failure to H-type aluminum alloy rod failure;
(2) As the wall thickness of the groove type connector increases from 4 mm to 14 mm, the yield displacement and yield load increase by 2.44 and 3.96 times, respectively, and the ultimate displacement and ultimate load increase by 1.8 and 2.46 times, respectively. The failure characteristic changes from the groove type connector failure to the H-type aluminum alloy rod failure;

(3) As the arch angle increases, there is a gradual increase in vertical bearing capacity of beam-column joints. As the wall thickness of groove connectors increases, the increase in vertical bearing capacity of beam-column joints is gradually reduced.

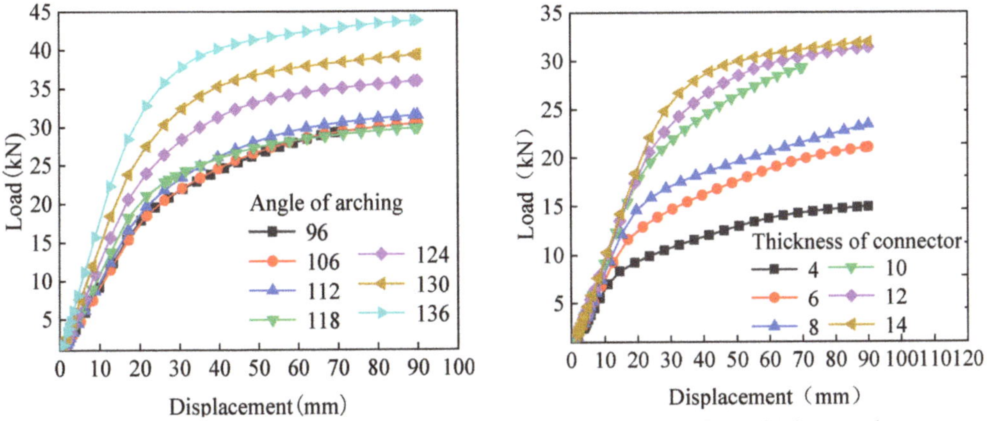

(**a**) The influence of arch angle (**b**) The influence of the thickness of connectors

Figure 14. Vertical load–displacement curve.

Table 4. Key results of different arch angles under vertical loads.

Arch Angle (°)	Yield Displacement (mm)	Yield Load (kN)	Failure Characteristics	Failure Displacement (mm)	Failure Load (kN)
96	8.38	5.94	Connector	48.38	26.10
106	8.21	7.47	Connector	45.71	24.53
112	8.81	8.67	Connector	44.15	27.07
118	8.21	8.94	Beam	48.85	29.25
124	7.25	9.75	Beam	48.65	33.10
130	8.21	11.98	Beam	44.21	36.10
136	7.25	12.13	Beam	46.65	40.14

Table 5. Key results of wall thickness of different connections under vertical load.

Thickness (mm)	Yield Displacement (mm)	Yield Load (kN)	Failure Characteristics	Failure Displacement (mm)	Failure Load (kN)
4	4.60	3.02	Connector	30.50	14.67
6	5.28	4.00	Connector	39.71	24.53
8	6.31	5.21	Connector	46.59	19.16
10	8.38	5.94	Beam	48.38	26.10
12	10.94	10.18	Beam	50.58	28.48
14	11.21	11.98	Beam	54.21	36.10

The yield stress state and ultimate stress state of aluminum alloy portal frame beam-column joints under vertical load at different arch angles are shown in Figure 15. When the beam-column joint begins to yield, as the arch angle increases, the stress of the H-type aluminum alloy rod gradually increases, and the stress of the bolt changes without obvious regularity. The stress of the groove type connection remains basically unchanged. When the beam-column joint reaches the limit state, with an increase in the arch angle, the stress of the H-type aluminum alloy beam and the groove connection remains basically unchanged, and the stress change in the bolts shows no obvious pattern.

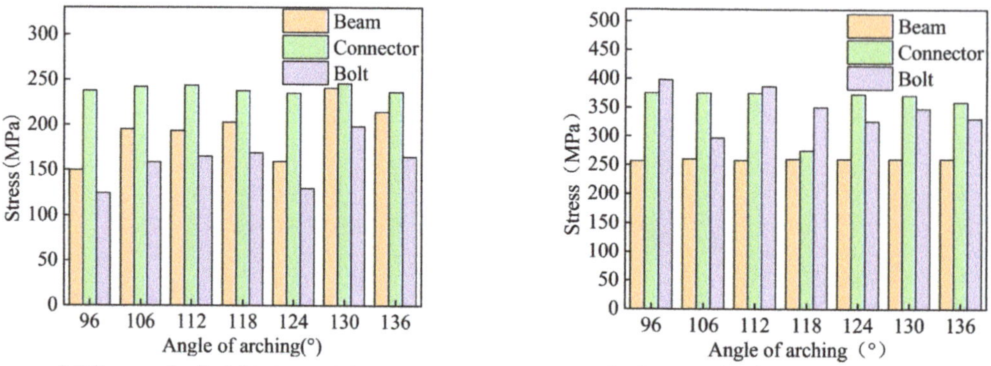

(**a**) Bar graph of yield stress state (**b**) Bar graph of ultimate stress state

Figure 15. Stress bar chart of different arch angles under vertical load.

The yield stress state and ultimate stress state of aluminum alloy portal frame beam-column joints under vertical load are shown in Figure 16, when the wall thickness of different groove connectors is different. As the wall thickness of the groove type connector increases, the stress of the H-type aluminum alloy rod gradually increases when the aluminum alloy portal frame beam-column joint yields. The stress of the groove type connector remains unchanged, and the stress of the bolt significantly increases. As the wall thickness of the groove type connector increases, when the beam-column joint reaches the limit state, the stress of the H-type aluminum alloy rod remains unchanged. The stress of the groove type connector has no obvious change pattern, and the change amplitude is small, and the stress of the bolt shows a trend of increasing first and then decreasing.

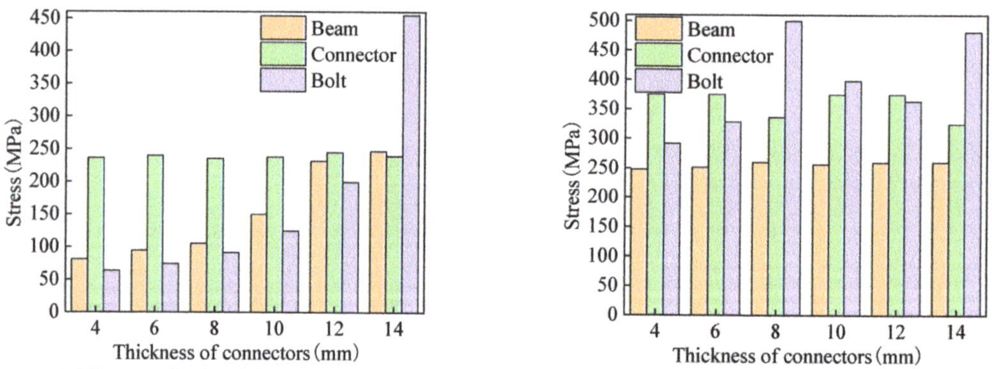

(**a**) Bar graph of yield stress state (**b**) Bar graph of ultimate stress state

Figure 16. Stress bar chart of different thicknesses of connectors under vertical load.

5.3. Mechanical Properties under Horizontal Loads

In aluminum alloy portal frames, beam-column joints mainly bear vertical and horizontal loads. The mechanical performance of this type of joint under vertical load has been previously explored, and now, we are conducting research on its mechanical performance under horizontal load. The influence of different bolt diameters, arch angles, and wall thicknesses of groove connectors on the load-bearing performance of aluminum alloy portal frame beam-column joints under horizontal load are summarized in Figure 17 and Tables 6–8. Through analysis, the conclusions drawn are as follows:

(1) Under horizontal load, as the bolt diameter increases, there is no significant change in the yield load, yield displacement, ultimate load, and ultimate displacement

of the beam-column joint, and the change amplitude is minimal. This is because when the bearing capacity of the bolt group is greater than the bearing capacity of the member section, increasing the bolt diameter does not enhance the horizontal bearing performance.

(2) When the beam-column joint is subjected to horizontal load, as the arch angle increases, the yield and ultimate displacement gradually increase (2.14 times and 2.78 times, respectively), and the yield and ultimate load gradually decrease (58% and 48%, respectively). This is because as the arch angle increases, the axial force generated by the horizontal load gradually decreases and the bending moment gradually increases.

(3) The wall thickness of different groove connectors has little effect on the horizontal load-bearing performance of aluminum alloy portal frame beam-column joints. For instance, as the wall thickness changes, the load–displacement curves basically overlap, and the changes in yield displacement, yield load, ultimate displacement, and ultimate load are minimal.

(4) Under horizontal loads, the failure mode of joints mainly manifests as beam failure and connector failure under different bolt diameters, arch angles, and connection wall thicknesses.

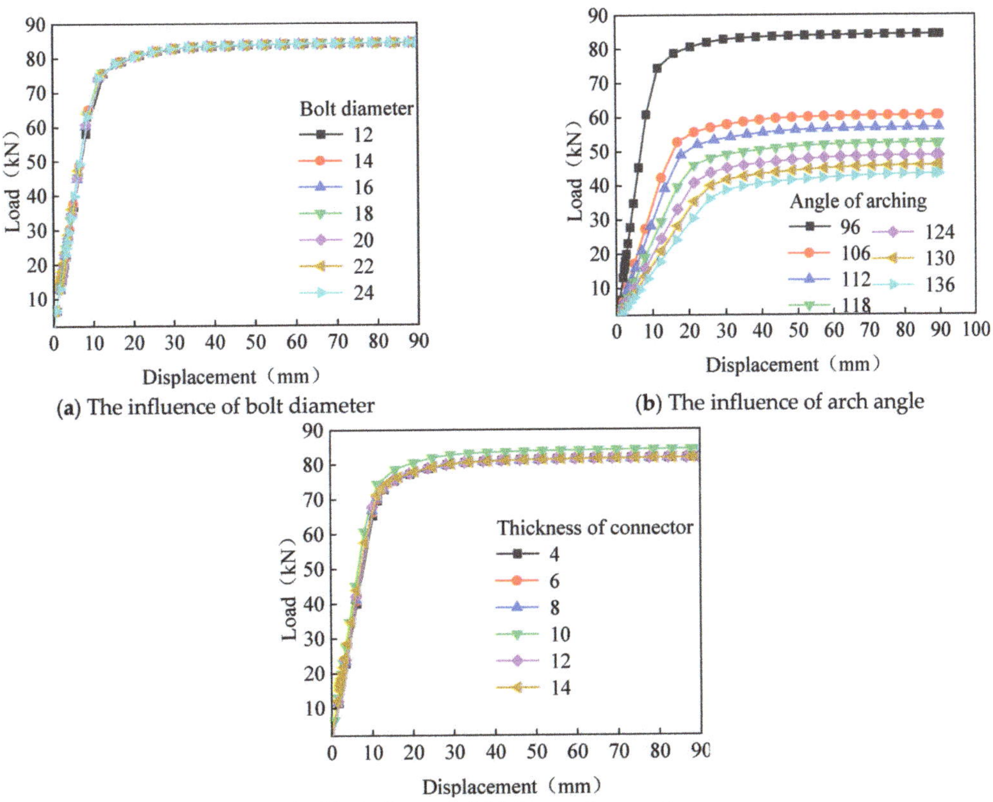

(**a**) The influence of bolt diameter (**b**) The influence of arch angle (**c**) The influence of thickness of connectors

Figure 17. Horizontal load–displacement curve.

Table 6. Key results of different bolt diameters under horizontal load.

Arch Angle (°)	Yield Displacement (mm)	Yield Load (kN)	Failure Characteristics	Failure Displacement (mm)	Failure Load (kN)
96	6.15	18.22	Beam	19.21	39.56
106	6.22	18.91	Beam	19.03	39.58
112	6.76	19.30	Beam	20.61	40.32
118	6.25	22.76	Beam	19.15	39.62
124	5.76	21.40	Beam	20.61	40.33
130	6.51	23.77	Beam	19.78	39.55
136	6.74	24.65	Beam	20.37	40.29

Table 7. Key results of different arch angles under horizontal loads.

Arch Angle (°)	Yield Displacement (mm)	Yield Load (kN)	Failure Characteristics	Failure Displacement (mm)	Failure Load (kN)
96	5.79	21.40	Beam	20.61	40.33
106	8.21	13.73	Beam	30.71	28.99
112	9.82	14.14	Beam	36.17	27.51
118	12.71	14.76	Beam	39.71	25.29
124	12.71	12.27	Beam	44.21	23.55
130	12.71	10.47	Beam	48.71	22.13
136	12.38	8.89	Beam	57.38	21.13

Table 8. Key results of wall thickness of different connections under horizontal load.

Thickness (mm)	Yield Displacement (mm)	Yield Load (kN)	Failure Characteristics	Failure Displacement (mm)	Failure Load (kN)
4	6.21	21.43	Connector	19.49	38.62
6	6.31	20.40	Connector	19.49	38.67
8	6.35	20.66	Connector	19.49	38.70
10	5.79	21.40	Beam	20.61	40.33
12	6.39	21.118	Beam	19.49	38.74
14	6.86	22.16	Beam	20.78	39.00

The yield and ultimate stress states of aluminum alloy portal frame beam-column joints under horizontal load with different bolt diameters are shown in Figure 18. There is no significant difference in the stress of H-type aluminum alloy rods at joint yield and ultimate state, while the stress of bolts at joint yield and ultimate state decreases with an increase in diameter. This is because an increase in bolt diameter will increase the stress area of the bolt, thereby significantly reducing its stress.

The yield and ultimate stress states of aluminum alloy portal frame beam-column joints under horizontal load at different arch angles are shown in Figure 19. As the arch angle increases, the stress of the H-shaped aluminum alloy beam remains unchanged during joint yielding and failure, the stress of the groove type connector increases first and then remains unchanged, and the stress of the bolts shows a trend of increasing first and then decreasing.

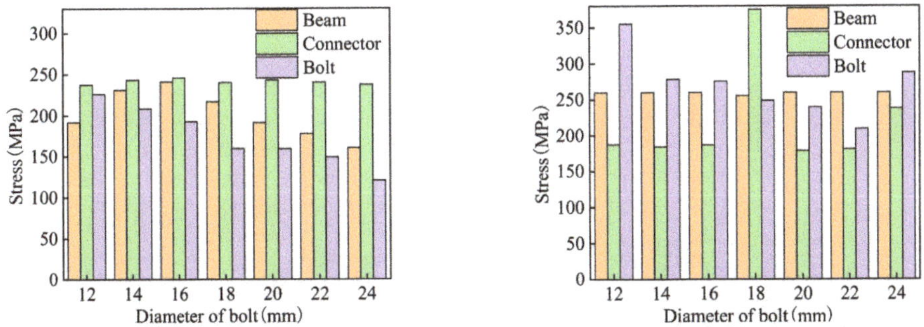

(**a**) Bar graph of yield stress state (**b**) Bar graph of ultimate stress state

Figure 18. Stress bar chart of different bolt diameters under horizontal load.

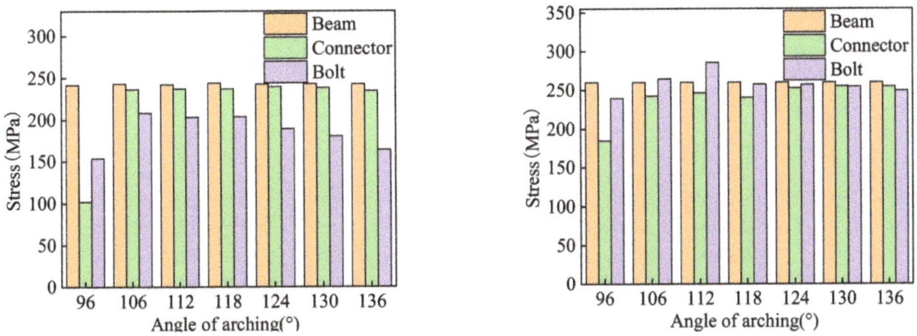

(**a**) Bar graph of yield stress state (**b**) Bar graph of ultimate stress state

Figure 19. Stress bar chart of different arch angles under horizontal load.

When the wall thickness of the groove type connector changes, the yield and ultimate stress state changes in the aluminum alloy portal frame beam-column joints under horizontal load are shown in Figure 20. As the wall thickness of the groove connection increases, the stress of the H-type aluminum alloy beam at joint yield gradually increases, but the increase is relatively small. The stress of the groove type connector has no obvious change pattern, and the change amplitude is relatively large when the joint yields, while the change amplitude is relatively small when the joint fails. When the joint yields and fails, the stress of the bolt increases with an increase in the wall thickness of the groove connection, and its variation is significant.

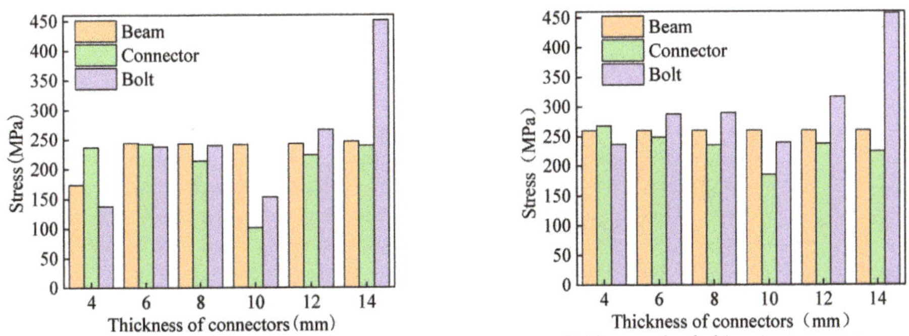

(**a**) Bar graph of yield stress state (**b**) Bar graph of ultimate stress state

Figure 20. Stress bar chart of different thicknesses of connectors under horizontal load.

6. Joint Construction Design

6.1. Joint Construction Improvement

Through experiments and numerical analysis of aluminum alloy portal frame beam-column joints, it can be concluded that the main failure feature of the joints is the failure of the web plate at the intersection of the beam-column joints with the groove connectors, as shown in Figure 21. For structures, the connection joints should not be damaged before the components, so it is necessary to take strengthening measures for these types of joints.

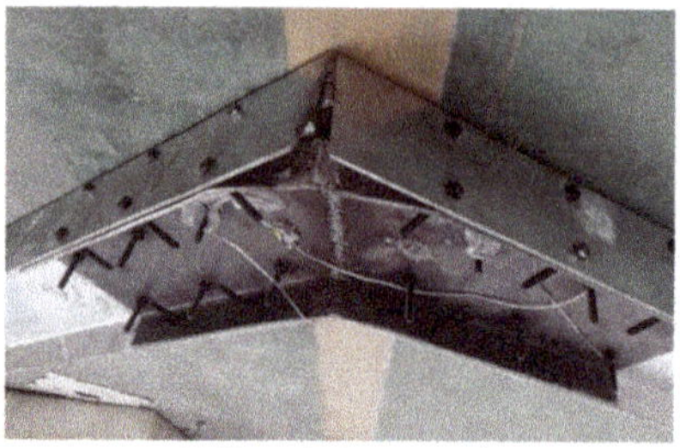

(**a**) Test failure mode

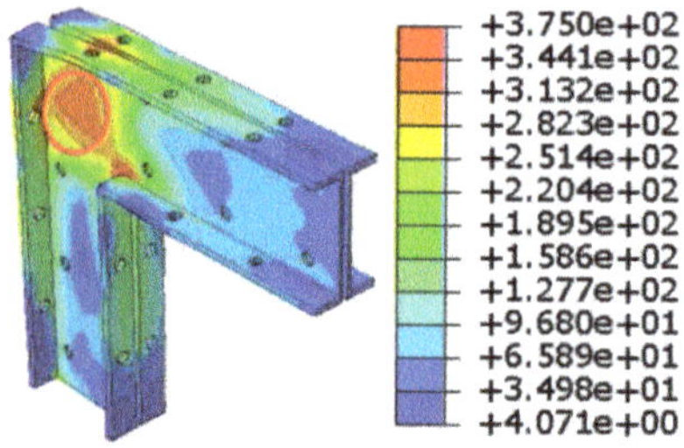

(**b**) Numerical simulation failure mode

Figure 21. Characteristics of joint structure failure.

According to the characteristics of the damage, structural strengthening measures such as adding partitions can be adopted, as shown in Figure 22. This partition can effectively improve the buckling resistance of the groove type connector's web plate (out-of-plane), while providing effective support for the flange of the groove type connector, thereby comprehensively improving the load-bearing performance of aluminum alloy portal frame beam-column joints.

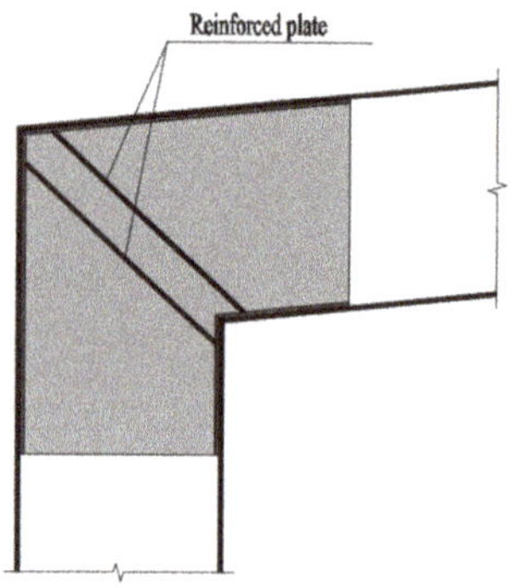

Figure 22. Strengthening measures for joint.

6.2. Joint Design Method

In order to meet the structural design principle of "strong joints and weak members", the joint design should meet the principle of not damaging the joint before the component. By analyzing the influence of different bolt diameters on the bearing performance of beam-column joints under vertical and horizontal loads, it can be concluded that when the bearing capacity of the bolt group is greater than that of aluminum alloy members, increasing the bolt diameter has no effect on the bearing performance of the joint [25,26]. Therefore, the bearing capacity of the bolt group only needs to be greater than that of the aluminum alloy rod:

$$n \min\left[d_w f_c \cdot 2t, \frac{\pi}{2} d^2 f_v\right] \geq f A_n \tag{1}$$

In the formula, n is the number of bolts, d is the diameter of the bolt, t is the thickness of the H-type aluminum alloy rod, f_c is the compressive strength of the aluminum alloy bolt connection, f_v is the shear strength of the bolt, f is the design strength of the aluminum alloy, and A_n is the net cross-sectional area of the H-type aluminum alloy rod.

The numerical simulation parameter analysis results for different wall thicknesses of groove connectors indicate that their thickness has a significant impact on the vertical bearing performance of beam-column joints, while their impact on the horizontal bearing performance is relatively small. The bearing capacity of the groove type connector should be greater than that of the H-type aluminum alloy rod. Therefore, the steel model of the groove type connector should ensure that its design strength is greater than the corresponding aluminum alloy rod, and the thicknesses of its web and flange should be greater than those of the aluminum alloy rod [26,27].

7. Conclusions

The aluminum alloy portal frame beam-column joint was considered as the research object, and experiments and numerical analysis were conducted to fully explore the mechanical performance of the joint under vertical and horizontal loads. The main conclusions are as follows:

(1) The experimental results indicate that as the diameter of the bolt increases, the damage situation at the bolt hole diameter of the specimen continues to improve, and the degree of compression at the bolt hole wall gradually decreases. The failure mode of beam-column joints does not change significantly with the change in arch angle, that is, the change in arch angle has no effect on the failure form and location of the specimen.

(2) The experimental results show that the load–displacement curve mainly consists of three stages, namely, the elastic stage, yield stage, and degradation stage. In the elastic stage, the load–displacement curves of different bolt diameters are basically consistent. After entering the plastic stage, the ultimate load first increases and then decreases, and the ultimate displacement is basically consistent.

(3) According to the experiment, there is no significant difference in the load–displacement curve when the arch angle increases from 90 degrees to 108 degrees. When the arch angle increases to 126 degrees, the stiffness and ultimate bearing capacity of the joint under vertical load significantly increase.

(4) As the wall thickness of the groove type connector increases from 4 mm to 14 mm, the vertical yield displacement and yield load increase by 2.44 and 3.96 times, respectively, and the ultimate displacement and ultimate load increase by 1.8 and 2.46 times, respectively.

(5) Under horizontal load, as the diameter of the bolt increases, the yield load, yield displacement, ultimate load, and ultimate displacement of the beam-column joint show no significant changes, and the amplitude of changes is minimal.

(6) When the beam-column joint is subjected to horizontal loads, as the arch angle increases, the yield and ultimate displacement increase by 2.14 times and 2.78 times, respectively, and the yield and ultimate load decrease by 58% and 48%, respectively.

(7) The research results of this article can be used for the design of aluminum alloy portal frames, and a series of studies on the mechanical performance of the overall portal frame structure should be carried out in the future.

Author Contributions: Conceptualization, G.W. and X.L.; methodology, Z.X.; software, Z.X.; validation, G.W., J.P. and C.Z.; formal analysis, Q.C.; investigation, G.W.; resources, X.L.; data curation, Z.X.; writing—original draft preparation, Z.X.; writing—review and editing, Z.X.; visualization, G.W.; supervision, X.L.; project administration, G.W.; funding acquisition, X.L. All authors have read and agreed to the published version of the manuscript.

Funding: This research was funded by [Research on a new type of aluminum alloy portal frame structure] grant number [YH2023001] And The APC was funded by [Guangzhou Yuehong Membrane Structure Engineering Co., Ltd.].

Data Availability Statement: Not applicable.

Acknowledgments: This research was financially supported by Guangzhou Yuehong Membrane Structure Engineering Co., Ltd., Guangzhou, China.

Conflicts of Interest: The authors declare no conflict of interest.

References

1. Guo, X.; Xiong, Z.; Luo, Y.; Xu, H.; Liang, S. Block tearing and local buckling of aluminum alloy gusset joint plates. *KSCE J. Civ. Eng.* **2016**, *20*, 820–831. [CrossRef]
2. Guo, X.; Xiong, Z.; Luo, Y.; Qiu, L.; Liu, J. Experimental investigation on the semi-rigid behaviour of aluminium alloy gusset joints. *Thin-Walled Struct.* **2015**, *87*, 30–40. [CrossRef]
3. Xiong, Z.; Guo, X.; Luo, Y.; Xu, H. Numerical analysis of aluminium alloy gusset joints subjected to bending moment and axial force. *Eng. Struct.* **2017**, *152*, 1–13. [CrossRef]
4. Guo, X.; Zhu, S.; Liu, X.; Liu, L. Experimental study on hysteretic behavior of aluminum alloy gusset joints. *Thin-Walled Struct.* **2018**, *131*, 883–901. [CrossRef]
5. Ma, H.; Yu, L.; Fan, F.; Yu, Z. Mechanical performance of an improved semi-rigid joint system under bending and axial forces for aluminum single-layer reticulated shells. *Thin-Walled Struct.* **2019**, *142*, 322–339. [CrossRef]
6. Xu, S.; Chen, Z.; Wang, X.; Mazzolani, F.M. Hysteretic out-of-plane behavior of the temcor joint. *Thin-Walled Struct.* **2015**, *94*, 585–592. [CrossRef]
7. Wu, Y.; Liu, H.; Chen, Z.; Liu, Y. Study on low-cycle fatigue performance of aluminum alloy temcor joints. *KSCE J. Civ. Eng.* **2020**, *24*, 195–207. [CrossRef]
8. Wang, G.; Zhao, C.Q.; Ma, J. Experimental and numerical study on the bending performance of an aluminium alloy flower-gusset composite joint. *Structures* **2021**, *33*, 2475–2486.
9. Wang, G.; Zhao, C.Q. Experimental and theoretical study on the bearing capacity of FGC joints for single-layer aluminium alloy lattice shell structures. *Structures* **2021**, *33*, 2445–2458.
10. Hiyama, Y.; Ishikawa, K.; Kato, S. Buckling behaviour of aluminum alloy double layer truss grid using ball joint system. In *Proceedings of the IASS-LAS98 Conference*; International Association for Shell and Spatial Structure (IASS): Madrid, Spain, 1998; Volume 9.
11. Liu, H.; Gu, A.; Chen, Z. Mechanical properties and design method of aluminum alloy bolt-sphere Joints. *Struct. Eng. Int.* **2019**, *31*, 30–39. [CrossRef]
12. Liu, H.; Gu, A.; Chen, Z. Tensile properties of aluminum alloy bolt-sphere joints under elevated temperatures. *KSCE J. Civ. Eng.* **2020**, *24*, 525–536. [CrossRef]
13. Shi, G.; Ban, H.; Bai, Y.; Wang, Y.; Luo, C.; Shi, Y. A novel cast aluminum joint for reticulated shell structures: Experimental study and modeling. *Adv. Struct. Eng.* **2013**, *16*, 1047–1059. [CrossRef]
14. Sugizaki, K.; Kohmura, S. Experimental study on buckling behaviour of a triodetic aluminum space frame. In Proceedings of the International Association for Shell and Space Structure, Tokyo, Japan, 19–22 October 1993; Volume 10, pp. 205–212.
15. Sugizaki, K.; Kohmura, S. Experimental study on buckling behaviour of a triodetic aluminum space frame: No.2 ultimate bearing strength of a single-layer space frame. In Proceedings of the IASS-ASCE International Symposium, Atlanta, GA, USA, 24–28 April 1994; Volume 4, pp. 478–484.
16. Yonemaru, K.; Fujisaki, T.; Nakatsuji, T. Development of space truss structure with CFRP. *Mater. Struct.* **1997**, *8*, 81–87.
17. Mazzolani, F.M. 3D aluminium structures. *Thin-Walled Struct.* **2012**, *61*, 258–266. [CrossRef]
18. Wang, Y.; Fan, F.; Lin, S. Experimental investigation on the stability of aluminium alloy 6082 circular tubes in axial compression. *Thin-Walled Struct.* **2015**, *89*, 54–66. [CrossRef]
19. *GB/T 50429-2007*; Code for Design of Aluminium Structures. General Administration of Quality Supervision, Inspection and Quarantine of the People's Republic of China: Beijing, China, 2007.
20. *GB/T 228.1-2010*; Metallic Materials—Tensile Testing—Part 1: Method of Test at Room Temperature. General Administration of Quality Supervision, Inspection and Quarantine of the People's Republic of China: Beijing, China, 2011.
21. Sun, Y.; Zhang, K.; Gong, G. Material properties of structural aluminium alloys after exposure to fire. *Structures* **2023**, *55*, 2105–2111. [CrossRef]
22. Ramberg, W.; Osgood, W.R. *Description of Stress-Strain Curves by Three Parameters*; NACA TN-902; National Advisory Committee for Aeronautics: Washington, DC, USA, 1943.
23. Sun, Y.; Fu, Z.; Song, Y.; Xia, J. Cross-Sectional Behavior of Aluminum Alloy Channel Section Stub Columns after Exposure to Fire. *J. Struct. Eng.* **2023**, *149*, 04023085. [CrossRef]
24. Zhou, W.; Ai, S.; Chen, M.; Zhang, R.; He, R.; Pei, Y.; Fang, D. Preparation and thermodynamic analysis of the porous ZrO_2/(ZrO_2 + Ni) functionally graded bolted joint. *Compos. Part B Eng.* **2015**, *82*, 13–22. [CrossRef]
25. Zhou, W.; Zhang, R.; Fang, D. Design and analysis of the porous ZrO_2/(ZrO_2+Ni) ceramic joint with load bearing-heat insulation integration. *Ceram. Int.* **2015**, *42*, 1416–1424. [CrossRef]

26. Tian, L.; Li, M.; Li, L.; Li, D.; Bai, C. Novel joint for improving the collapse resistance of steel frame structures in column-loss scenarios. *Thin-Walled Struct.* **2023**, *182*, 110219. [CrossRef]
27. Yang, Y.; Lin, B.; Zhang, W. Experimental and numerical investigation of an arch–beam joint for an arch bridge. *Arch. Civ. Mech. Eng.* **2023**, *23*, 101. [CrossRef]

buildings

MDPI

Article

Impact of Tunneling on Adjacent Piles Based on the Kerr Foundation Model Considering the Influence of Lateral Soil

Haipeng Jia [1,2], Ning Wang [3,4,*], Haibin Ding [4] and Lingxiao Guan [4]

1 School of Civil Engineering, Tianjin University, Tianjin 300072, China; 13728759267@163.com
2 Shenzhen Geotechnical Investigation & Surveying Institute (Group) Co., Ltd., Shenzhen 518028, China
3 School of Civil Engineering, Central South University of Forestry and Technology, Changsha 410004, China
4 State Key Laboratory for Performance Monitoring and Guarantee of Rail Transportation Infrastructure, East China Jiaotong University, Nanchang 330013, China; hbding@ecjtu.edu.cn (H.D.); glx1392@163.com (L.G.)
* Correspondence: wn1290@163.com

Abstract: The Kerr foundation model simulates the interaction between piles and soil. Considering the impact of lateral soil displacement on adjacent piles, the lateral displacement and bending moment of the adjacent piles caused by shield tunnel excavation are calculated in detail. Additionally, the reactions of groups of piles are obtained by focusing on the shielding effect of the piles on the soil displacement caused by shield tunnel excavation. The validity of the solutions is verified by comparing the calculated results with the boundary element program GEPAN. Additionally, adjacent pile lateral displacement and bending moment are compared, with and without considering lateral soil effects. Furthermore, this study investigates the influence of various factors, such as soil spring stiffness, pile–tunnel distance, ground loss ratio, and pile diameter on the pile group's lateral displacement and bending moment. The research findings indicate that increasing the soil spring stiffness or the horizontal distance between the pile and tunnel can reduce the lateral displacement and the bending moment of the pile. On the other hand, as the ground loss ratio gradually increases, the pile lateral displacement and bending moment will also increase. However, when the diameter of the pile grows, the lateral displacement reduces, while the bending moment increases.

Keywords: tunnel–soil–pile interaction; Kerr foundation model; lateral soil displacements; shielding effect; group piles

Citation: Jia, H.; Wang, N.; Ding, H.; Guan, L. Impact of Tunneling on Adjacent Piles Based on the Kerr Foundation Model Considering the Influence of Lateral Soil. *Buildings* **2023**, *13*, 2548. https://doi.org/10.3390/buildings13102548

Academic Editor: Giuseppina Uva

Received: 17 August 2023
Revised: 4 October 2023
Accepted: 7 October 2023
Published: 9 October 2023

1. Introduction

In densely populated urban areas, metro tunnels are frequently constructed near pre-existing pile foundations, which could adversely affect their safety and serviceability. Tunnel excavations can relax ground pressure and cause soil movements. This can lead to additional deformations and forces in neighboring piles [1]. These effects pose a significant challenge to engineers involved in the design and construction of tunnels.

There are three main categories of methods used to investigate the effects of tunneling on existing piles, which are field monitoring and centrifuge model tests, complete numerical analyses, and simplified two-stage procedures. Investigating tunnel-induced effects on nearby ground and structures is often conducted using field monitoring, a straightforward and reliable method. Boonyarak et al. [2] recorded various ground and pier responses during tunnel driving. Based on these field observation data, they found that it is not feasible to instrument a pile buried underground before tunneling. Alternatively, many researchers have used centrifuge model tests to investigate pile behavior under various tunnel layouts and excavation procedures [3–9]. Additionally, a complete numerical analysis is critical for evaluating the impacts of tunneling on adjacent ground and piles. The finite element method is the most widely used for comprehensive numerical analysis. In previous studies [10–15], the pile–soil–tunnel system is treated as an integral unit, and the nonlinear behavior of the soil was also simulated, as well as the complex construction sequence.

Researchers have dedicated significant efforts to experimental and numerical studies. However, these methods necessitate extensive computational time or specialized equipment. Thus, a simplified two-stage process was developed to provide preliminary design engineers with a convenient reference. The first stage involves estimating free-field soil movements caused by tunneling without the presence of piles, while in the second stage these movements are imposed on the piles to calculate their response. Chen et al. [16] analyzed a single pile's lateral and axial responses by combining the analytical soil movements proposed by Loganathan [17] with a simplified boundary element method. Loganathan et al. [18] investigated the behavior of a group of piles with various configurations and subjected to ground movement induced by tunneling in the same way. Huang and colleagues [19] employed the Winkler foundation model in the second stage to simulate pile–soil interactions and study the response of a passive pile group to tunneling. Mu and coworkers [20] researched pile raft vertical and lateral responses to ground movements resulting from tunnel excavation in layered soil. The Winkler foundation model has been widely accepted due to its simple nature. However, it disregards the continuity and shear characteristics of the adjacent soil layers. To address this issue, the Pasternak foundation model was introduced [21]. This model adds a shear layer to the Winkler model, connecting the adjacent spring elements. Zhang et al. [22] used the Pasternak model to study the lateral response of single piles and group piles to tunneling. Furthermore, most current analytical two-stage studies rely on the plane strain assumption. In practice, the soil movement induced by tunneling varies within a specific range along the direction of tunnel excavation and shows apparent 3D deformation. Zhang et al. [22] also considered the 3D effect of lateral soil displacement in their work. Kerr [23] modified the Pasternak model by including an upper spring layer, which prevents mathematical difficulties such as infinite reaction pressures at the foundation's edge and provides an extra boundary condition to restrict the foundation. Zhang et al. [24] integrated a cavity contraction theory and the Kerr foundation model to suggest an analytical solution for forecasting tunneling-induced distortions for a single pile. In summary, studies based on the Kerr model and considering the lateral soil displacement effect have yet to take place for the response of group piles to tunneling.

This paper uses the two-stage method and the Kerr foundation model to investigate the deflections and internal forces of existing single piles and group piles caused by tunnel excavation. The paper also considers the effect of lateral soil displacement near the pile foundations. The computational results were compared with and without considering the impact of lateral soil displacement adjacent to the piles. The validity of the solution was verified by comparing the predicted results with the boundary element program GEPAN. Finally, a parametric analysis was performed to analyze the effects of various factors on the lateral displacement and bending moment of the group piles, including soil spring stiffness, pile–tunnel distance, ground loss rate, and pile diameter.

2. Pile Response without Considering the Effects of Lateral Soil

2.1. Lateral Displacement of Soil

According to the linear elasticity theory, Loganathan et al. [17] proposed an analytical solution to determine the vertical displacement of the adjacent soil free field caused by shield tunnel excavation in 1998. The calculated results showed good agreement with the measured values. Therefore, this paper employs the previously mentioned formulas to determine the lateral displacement of the adjacent soil mass caused by shield tunnel excavation. The following expression is used:

$$U(z) = -R^2 x_0 \varepsilon_0 e^{-\left[\frac{1.38{x_0}^2}{(H+R)^2}+\frac{0.69z^2}{H^2}\right]} \cdot \left\{ \frac{1}{x_0^2+(z-H)^2} - \frac{4z(z+H)}{\left[x_0^2+(z+H)^2\right]^2} + \frac{(3-4v)}{x_0^2+(z+H)^2} \right\} \quad (1)$$

where R is the radius of the shield tunnel, x_0 is the lateral distance between the pile foundation and tunnel center line, ε_0 is the average loss rate of surrounding soil caused by shield tunnel excavation, H is the burial depth of the tunnel axis, z is the vertical distance from the ground surface, and v is the Poisson ratio of the soil.

2.2. Lateral Response Analysis of a Single Pile in Kerr Foundation

The Pasternak foundation model adds a variable elastic deformation coefficient to the Kerr foundation model. In other words, the Kerr foundation model contains two spring layers and one shear layer of soil, as shown in Figure 1. k, c, and g are the three parameters corresponding to the Kerr foundation model. p(z), w(z), and w_2(z) denote the additional load resulting from shield tunnel excavation, the lateral displacement of the pile foundation, and the shear layer of soil, respectively.

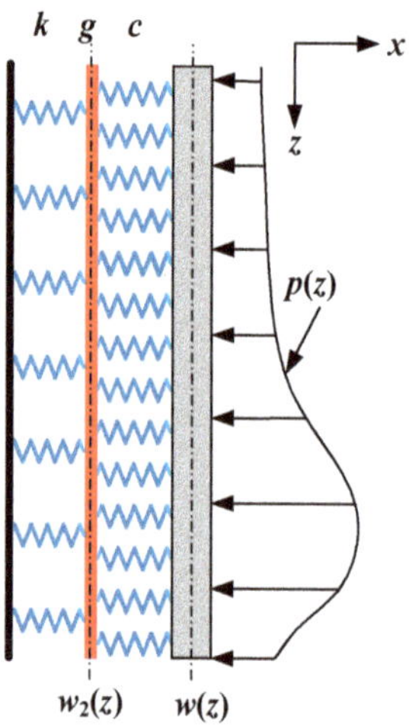

Figure 1. Pile–soil interaction model in Kerr's foundation.

According to the pile–soil interaction model of the Kerr foundation, the following assumptions are considered:

- The pile foundation is modeled as a cylindrical beam that rests on the Kerr foundation;
- The shear layer in the Kerr foundation model only experiences shear deformation;
- The pile foundation maintains constant contact with the foundation soil, with the two exhibiting well-coordinated deformation at the interface;
- Lateral friction between the foundation and pile foundation is not accounted for.

The additional load p(z) generated by the excavation of a shield tunnel on a pile foundation should also be satisfied as the following equation:

$$\left(1+\frac{k}{c}\right)p(z)-\frac{g}{c}\frac{\mathrm{d}^2p(z)}{\mathrm{d}z^2}=kU(z)-g\frac{\mathrm{d}^2U(z)}{\mathrm{d}z^2} \tag{2}$$

in which c and k independently represent the stiffness of the first-layer and the second-layer soil spring [25], whose relationship is assumed as $c = nk$; g is the shear layer's stiffness; and k and g can be calculated as [26]:

$$k=\frac{0.65E_s}{D(1-v^2)}\left(\frac{E_sD^4}{EI}\right)^{1/12} \tag{3}$$

$$g=\frac{E_st}{6(1+v)} \tag{4}$$

where t, E_s, and ν characterize the thickness of the shear layer, the elastic modulus, and the Poisson ratio of soil, respectively. D, E, and I denote the diameter of the pile foundation, the elastic modulus, and the inertia moment of the pipeline. The term EI represents flexural stiffness.

According to Equation (2), the additional load $p(z)$ can be obtained by eliminating the higher-order approximation term as

$$p(z) = \eta\left[\frac{nk}{n+1}U(z) - \frac{n^2 g}{(n+1)^2}\frac{\mathrm{d}^2 U(z)}{\mathrm{d}z^2}\right] \tag{5}$$

in which η is the correction coefficient of load, and its value range is 0.4~1.0. Assuming that the pile on the Kerr foundation has a lateral displacement $w(z)$ under the action of this load, it can be written as

$$w(z) = w_1(z) + w_2(z) \tag{6}$$

where $w_1(z)$ and $w_2(z)$ are the deformation of the spring and shear layer of the soil on the right side. Supposing that the stress of the pile and shear layer on the left side, respectively, are expressed as

$$q_1(z) = cw_1(z) = c[w(z) - w_2(z)] \tag{7}$$

$$q_2(z) = kw_1(z) \tag{8}$$

Concerning the shear layer, there exists

$$q_1(z) = -g\frac{\mathrm{d}^2 w_2(z)}{\mathrm{d}z^2} + kw_2(z) \tag{9}$$

In combination with Equations (7) and (9), we acquire

$$w(z) = \left(1 + \frac{k}{c}\right)w_2(z) - \frac{g}{c}\frac{\mathrm{d}^2 w_2(z)}{\mathrm{d}z^2} \tag{10}$$

The governing equation of a pile foundation's displacement caused by shield tunnel excavation can be written as follows:

$$EI\frac{\mathrm{d}^4 w(z)}{\mathrm{d}z^4} + q_1(z)D = p(z)D \tag{11}$$

Substituting Equations (9) and (10) into Equation (11), the following equation can be gained:

$$\frac{EIg}{Dc}\frac{\mathrm{d}^6 w_2(z)}{\mathrm{d}z^6} - \frac{EI(c+k)}{Dc}\frac{\mathrm{d}^4 w_2(z)}{\mathrm{d}z^4} + g\frac{\mathrm{d}^2 w_2(z)}{\mathrm{d}z^2} - kw_2(z) = -p(z) \tag{12}$$

and Equation (12) can be given as the finite difference form:

$$\alpha(w_2)_{i+3} + \beta(w_2)_{i+2} + \chi(w_2)_{i+1} + \delta(w_2)_i + \chi(w_2)_{i-1} + \beta(w_2)_{i-2} + \alpha(w_2)_{i-3} = -p_i \tag{13}$$

where $\begin{pmatrix}\alpha\\ \beta\\ \chi\\ \delta\end{pmatrix} = \begin{pmatrix}1 & 0 & 0 & 0\\ -6 & -1 & 0 & 0\\ 15 & 4 & 1 & 0\\ -20 & -6 & -2 & -1\end{pmatrix}\cdot\begin{pmatrix}\frac{EIg}{Dcl^6}\\ \frac{EI(c+k)}{Dcl^4}\\ \frac{g}{l^2}\\ k\end{pmatrix}$, $l = L/n$ refers to the length of the pile foundation divided into n segments with length L, and $i = 0, 1, 2, \ldots, (n-1), n$. For brevity of calculation, three virtual nodes are added at the top and end of the pile, corresponding to nodes $-3, -2, -1, n+1, n+2$, and $n+3$ in turn.

Invoking the boundary conditions and eliminating the nodes above, the equation of a single pile's displacement can be solved as follows:

$$\{W_2\} = \{K\}^{-1} \cdot \{P\} \tag{14}$$

in which

$$\begin{aligned} \{W_2\} &= \left[(w_2)_0, (w_2)_1, \cdots, (w_2)_{n-1}, (w_2)_n\right]^{\mathrm{T}} \\ \{P\} &= \left[-p_0, -p_1, \cdots, -p_{n-1}, -p_n\right]^{\mathrm{T}} \end{aligned}$$

When there is no constraint at the top and tip of the pile foundation, the bending moment M_s of the soil's shear layer, the shear force Q_p, and the bending moment M_p for the exact position of the pile foundation are zero, that is

$$\begin{cases} M_{p0} = M_{pn} = -EI \dfrac{\mathrm{d}^2 w(z)}{\mathrm{d}z^2}\bigg|_{i=0,i=n} = 0 \\ M_{s0} = M_{sn} = -EI \dfrac{\mathrm{d}^2 w_2(z)}{\mathrm{d}z^2}\bigg|_{i=0,i=n} = 0 \\ Q_{p0} = Q_{pn} = -EI \dfrac{\mathrm{d}^3 w_2(z)}{\mathrm{d}z^3}\bigg|_{i=0,i=n} = 0 \end{cases} \tag{15}$$

in which

$$\begin{cases} \dfrac{\mathrm{d}^2 w_2(z)}{\mathrm{d}z^2} = \dfrac{(w_2)_{i+1} - 2(w_2)_i + (w_2)_{i-1}}{l^2} \\ \dfrac{\mathrm{d}^3 w_2(z)}{\mathrm{d}z^3} = \dfrac{(w_2)_{i+2} - 2(w_2)_{i+1} + 2(w_2)_{i-1} - (w_2)_{i-2}}{2l^3} \end{cases}$$

Therefore, the lateral stiffness matrix of the pile **K** can be obtained as

$$\mathbf{K} = \begin{bmatrix} A_1 & A_3 & A_6 & 2\alpha & & & & & \\ A_2 & A_5 & A_7 & \beta & \alpha & & & & \\ A_4 & A_7 & \delta & \chi & \beta & \alpha & & & \\ \alpha & \beta & \chi & \delta & \chi & \beta & \alpha & & \\ & \cdot & \cdot & \cdot & \cdot & \cdot & \cdot & \cdot & \\ & \cdot & \cdot & \cdot & \cdot & \cdot & \cdot & \cdot & \\ & \cdot & \cdot & \cdot & \cdot & \cdot & \cdot & \cdot & \\ & & \alpha & \beta & \chi & \delta & \chi & \beta & \alpha \\ & & & \alpha & \beta & \chi & \delta & A_7 & A_4 \\ & & & & \alpha & \beta & A_7 & A_5 & A_2 \\ & & & & & 2\alpha & A_6 & A_3 & A_1 \end{bmatrix} \tag{16}$$

where $A_1 = \delta + 2\chi + 4\beta + 8\alpha$, $A_2 = \chi + 2\beta + 2\alpha$, $A_3 = -4\beta - 10\alpha$, $A_4 = \beta + 2\alpha$, $A_5 = \delta - \beta$, $A_6 = 2\beta + 2\alpha$, $A_7 = \chi - \alpha$. After $\mathbf{W_2}$ is obtained, the pile foundation's lateral displacement can be worked out by associating with Equation (7).

2.3. Lateral Response Analysis of Group Piles in Kerr Foundation

Due to the shielding effect of pile foundations on the displacement of soil, not only the influence of shield tunnel excavation on adjacent group piles should be considered but also the mutual interaction between adjacent pile foundations. Figure 2 depicts a simplified model diagram for calculating the influence of shield tunnel excavation on adjacent group piles. x_1 and x_2 are the horizontal distance between two piles and the tunnel axis, as shown in Figure 2, respectively, and s denotes the axial distance between pile I (front) and pile II (back).

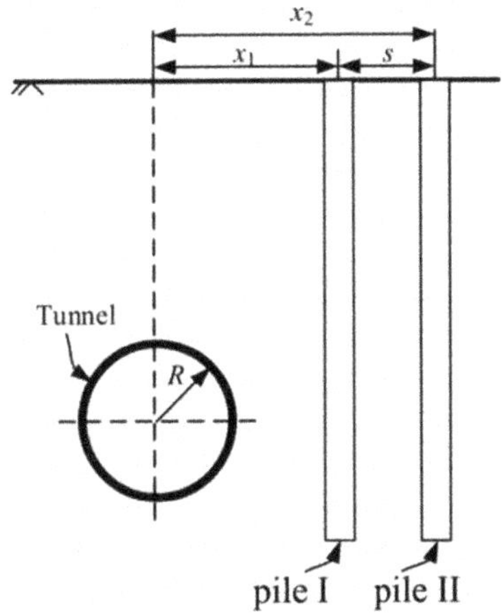

Figure 2. Computational model for group piles adjacent to tunneling.

The lateral soil displacement caused by shield tunnel excavation at the position of pile I is $U_1(z)$, which simultaneously results in the horizontal deflection of pile I, $W_1(z)$. Therefore, the shielding displacement due to the shielding effect of pile I can be described as $\Delta W_1(z) = U_1(z) - W_1(z)$. According to the transfer coefficient of soil displacement between piles proposed in the literature [27], the soil displacement caused by the shielding effect of pile I at the position of pile II can be obtained:

$$U_{21}(z) = \psi(s)\Delta W_1(z) = U_1(z) - W_1(z) \tag{17}$$

where $\psi(s) = \begin{cases} \frac{\ln(r_m)-\ln(s)}{\ln(r_m)-\ln(r_0)} & , r_0 \le s \le r_m \\ 0 & , s > r_m \end{cases}$. r_m is the effective influence radius of the pile, which can be expressed as $r_m = \chi_1\chi_2 L(1-v)$. Both χ_1 and χ_2 are the empirical coefficients. In Kerr's foundation, the additional load of pile II aroused from the shielding effect of pile I on the soil is

$$p_{21}(z) = \eta\left[\frac{nk}{n+1}U_{21}(z) - \frac{n^2 g}{(n+1)^2}\frac{d^2U_{21}(z)}{dz^2}\right] \tag{18}$$

Subsequently, the horizontal displacement $W_{21}(z)$ of pile II caused by the shielding effect of pile I on soil can be recovered by substituting Equation (18) into Equations (10) and (14). The horizontal displacement of pile i due to shielding tunnel excavation is expressed as $W_i(z)$. $W_{ij}(z)$ represents the horizontal displacement of pile i due to the shielding effect of pile j on soil. The horizontal displacement $W_{ii}(z)$ of pile i under the joint influence of shield tunnel excavation and adjacent piles is characterized as

$$W_{ii}(z) = W_i(z) + \sum_{j=1}^{m} W_{ji}(z)\,(j \ne i) \tag{19}$$

3. Lateral Displacement of Pile Foundation Considering the Effect of Lateral Soil

3.1. Lateral Response Analysis of a Single Pile in the Kerr Foundation

As shown in Figure 3a, the additional loads generated by the shield tunnel excavation not only act on the pile foundation but also significantly affect the soil next to the piles, parallel to the tunnel axis. The soil surrounding the piles also affects the displacement of the pile foundation, due to the inconsistent deformation of the ground and the pile foundation, as shown in Figure 3b.

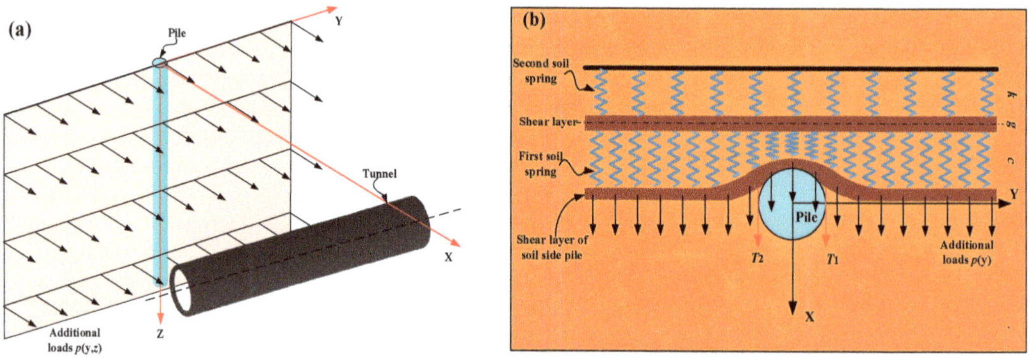

Figure 3. Influence of tunneling on the adjacent pile: (**a**) additional loads applied on the pile and soil, (**b**) forces acting on the pile through the lateral shear layer of soil.

When the Kerr foundation model is adopted to analyze the influence of soil on a pile foundation, fundamental assumptions are made about the following:

- The parameters of lateral soil parallel to the tunnel axis beside the pile foundation are consistent with the foundation soil;
- The lateral forces on the pile are T_1 and T_2, which are transferred to both sides of the pile through the shear layer of soil;
- The pile foundation is always in close contact with the soil next to the pile, and the deformation of the pile foundation is consistent with that of the soil's shear layer around the pile on the Kerr foundation;
- The additional load caused by shield tunnel excavation acts on the pile foundation and lateral soil in the meantime, supposing that the load's influence range is wide enough.

For any plane $z = z_0$, the governing equation of deformation for the soil's shear layer beside the pile under the action of additional load p is

$$p(y) = k\overline{U}(y) - g\frac{\mathrm{d}^2\overline{U}(y)}{\mathrm{d}y^2} \tag{20}$$

where $\overline{U}(y)$ is the deformation of the soil's shear layer around the pile along the y-axis direction. When $y \geq D/2$, the general solution of Equation (20) can be obtained as follows:

$$\overline{U}(y) = C_1 e^{-\sqrt{k/g}(y-D/2)} \tag{21}$$

As shown in Figure 4, $w_\mathrm{u}(z)$ is the particular solution of Equation (21) when $y = y_0$. The plane of $y = y_0$ is outside the effective influence range of the pile foundation. Hence, we can calculate Equation (20) as

$$\overline{U}(z\,,y) = C_1 e^{-\sqrt{k/g}(y-D/2)} + w_u(z) \tag{22}$$

Assuming that the pile foundation is always in contact with the soil adjacent to the pile, the deformation of the shear layer of the soil around the pile, located at $y = D/2$, is similar to the shear layer deformation (denoted as w_2) of soil in the Kerr foundation. Therefore, it can be solved by following the equation below:

$$C_1 = \overline{U} - w_u = w_2 - w_u \tag{23}$$

$$\overline{U}(z,y) = [w_2(z) - w_u(z)]e^{-\sqrt{k/g}(y-D/2)} + w_u(z) \tag{24}$$

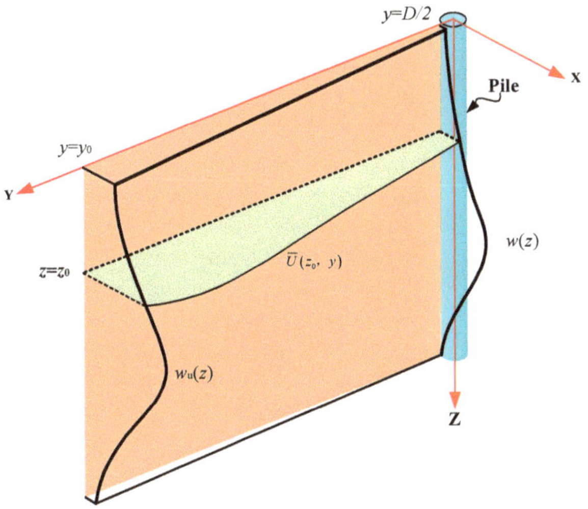

Figure 4. Deformation of the shear soil layer beside the pile.

Then, the force on the pile foundation can be expressed as

$$T_1(z) = T_2(z) = g\frac{\mathrm{d}\overline{U}(z,y)}{\mathrm{d}y} = [w_u(z) - w_2(z)]\sqrt{kg} \tag{25}$$

in which $w_\mathrm{u}(z)$ is the deformation of the soil's shear layer caused by the additional load on the $y = y_0$ plane due to shield tunnel excavation, and then the governing equation of $w_\mathrm{u}(z)$ can be obtained as

$$p(z) = \eta\left[\frac{nk}{n+1}w_u(z) - \frac{n^2 g}{(n+1)^2}\frac{\mathrm{d}^2 w_u(z)}{\mathrm{d}z^2}\right] \tag{26}$$

The governing equation of a pile foundation's deformation considering the influence of soil besides the pile on the pile foundation can be written as follows:

$$EI\frac{\mathrm{d}^4 w(z)}{\mathrm{d}z^4} + q_1(z)D = 2T_1(z) \tag{27}$$

Substituting Equations (9), (10), and (25) into Equation (27) can generate the following equations:

$$\frac{EI}{2c}\sqrt{\frac{g}{k}}\frac{\mathrm{d}^6 w_2(z)}{\mathrm{d}z^6} - \left[\frac{EI(c+k)}{2c\sqrt{gk}}\right]\frac{\mathrm{d}^4 w_2(z)}{\mathrm{d}z^4} + \left(\frac{D}{2}\sqrt{\frac{g}{k}}\right)\frac{\mathrm{d}^2 w_2(z)}{\mathrm{d}z^2} - \left(\frac{D}{2}\sqrt{\frac{k}{g}} + 1\right)w_2(z) = -w_u(z) \tag{28}$$

In addition, Equation (28) can also be expressed using the finite difference form, as follows:

$$\gamma_1(w_2)_{i+3} + \gamma_2(w_2)_{i+2} + \gamma_3(w_2)_{i+1} + \gamma_4(w_2)_i + \gamma_3(w_2)_{i-1} + \gamma_2(w_2)_{i-2} + \gamma_1(w_2)_{i-3} = -(w_u)_i \tag{29}$$

and

$$\begin{pmatrix}\gamma_1\\ \gamma_2\\ \gamma_3\\ \gamma_4\end{pmatrix} = \begin{pmatrix}1 & 0 & 0 & 0\\ -6 & -1 & 0 & 0\\ 15 & 4 & 1 & 0\\ -20 & -6 & -2 & -1\end{pmatrix}\cdot\begin{pmatrix}\frac{EI}{2cl^6}\sqrt{\frac{g}{k}}\\ \frac{EI(c+k)}{2c\sqrt{gkl^4}}\\ \frac{D}{2l^2}\sqrt{\frac{g}{k}}\\ \frac{D}{2}\sqrt{\frac{k}{g}} + 1\end{pmatrix} \tag{30}$$

Based on the unconstrained boundary conditions at the top and tip of the pile foundation and eliminating the nodes of −3, −2, −1 and $n + 1, n + 2, n + 3$, the governing equation of a single pile can be given as

$$\{W_2\} = \{K_u\}^{-1} \cdot \{-w_u\} \tag{31}$$

where $\{-w_u\} = \left[(-w_u)_0, (-w_u)_1, \ldots, (-w_u)_{n-1}, (-w_u)_n\right]^{\mathrm{T}}$, $\mathbf{K_u}$ denotes the lateral stiffness matrix of the pile in the x-direction of soil, which can be expressed as

$$\mathbf{K_u} = \begin{bmatrix} B_1 & B_3 & B_6 & 2\gamma_1 & & & & & \\ B_2 & B_5 & B_7 & \gamma_2 & \gamma_1 & & & & \\ B_4 & B_7 & \gamma_4 & \gamma_3 & \gamma_2 & \gamma_1 & & & \\ \gamma_1 & \gamma_2 & \gamma_3 & \gamma_4 & \gamma_3 & \gamma_2 & \gamma_1 & & \\ & \cdot & \cdot & \cdot & \cdot & \cdot & \cdot & \cdot & \\ & \cdot & \cdot & \cdot & \cdot & \cdot & \cdot & \cdot & \\ & \cdot & \cdot & \cdot & \cdot & \cdot & \cdot & \cdot & \\ & & \gamma_1 & \gamma_2 & \gamma_3 & \gamma_4 & \gamma_3 & \gamma_2 & \gamma_1 \\ & & & \gamma_1 & \gamma_2 & \gamma_3 & \gamma_4 & B_7 & B_4 \\ & & & & \gamma_1 & \gamma_2 & B_7 & B_5 & B_2 \\ & & & & & 2\gamma_1 & B_6 & B_3 & B_1 \end{bmatrix} \tag{32}$$

in which $B_1 = \gamma_4 + 2\gamma_3 + 4\gamma_2 + 8\gamma_1$, $B_2 = \gamma_3 + 2\gamma_2 + 2\gamma_1$, $B_3 = -4\gamma_4 - 10\gamma_1$, $B_4 = \gamma_2 + 2\gamma_1$, $B_5 = \gamma_4 - \gamma_2$, $A_6 = 2\gamma_2 + 2\gamma_1$, $A_7 = \gamma_3 - \gamma_1$. Meanwhile, the lateral deformation W_2 of the soil's shear layer, considering the effect of soil beside the pile in the Kerr foundation model, can be solved. The lateral displacement **W** of the pile foundation caused by shield tunnel excavation can be obtained by combining with Equation (7).

3.2. *Lateral Response Analysis of Group Piles in Kerr Foundation*

As a result, the same procedure solution as in Section 2.2 can be adapted to work out the lateral displacement of group piles in the Kerr foundation caused by shield tunnel excavation, considering the effect of lateral soil, as follows:

$$W_{ii}(z) = W_i(z) + \sum_{j=1}^{m} W_{ji}(z)\ (j \neq i) \tag{33}$$

4. Verifications of the Analytical Solution

The 3D boundary element program GEPAN, which considers the three-dimensional effect of lateral soil in the solution process, was utilized to verify the accuracy of the method proposed in this paper. Loganathan [18] simulated the impact of shield tunnel excavation on the horizontal displacement of adjacent 2 × 2 group piles using the 3D boundary element program GEPAN. The soil was assumed to be a homogeneous and isotropic elastic medium with a constant soil strength of C_u = 60 kPa in the model above. The soil modulus was E_s = 24 MPa. Piles were cylindrical structures made of uniform, isotropic elastic material with fixed elastic parameters. In the computer program GEPAN, a typical pile element discretization established the pile model. Both the piles and the surrounding soil mass deformed elastically, and the principle of superposition was applicable. Moreover, there was no slip between the piles and the soil at the interface. The distance between the front pile, the back pile, and the tunnel axis were 4.5 m and 6.9 m, respectively, as shown in Figure 5. The axial distance between the two piles was s = 2.4 m. The remaining calculation parameters related to the method in this paper are shown in Table 1.

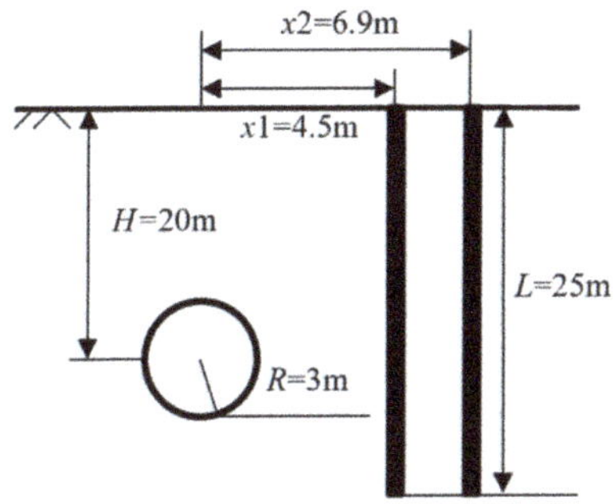

Figure 5. Calculation model of the influence of tunneling on group piles.

Table 1. Calculation parameters.

R (m)	H (m)	ε_0 (%)	E_s (MPa)	v	E_p (MPa)	D (m)	L (m)
3	20	1	24	0.5	3×10^4	0.8	25

Figure 6 shows the lateral displacement of the soil at the center positions of the front and back piles using the two-stage method. It is clear from the figure that the maximum lateral displacement of the soil was reduced significantly as the horizontal distance increased. Figures 7 and 8 show comparisons of lateral displacement and bending moment of the front pile and the back pile obtained using the simplified calculation method in this work with those from GEPAN [18] and the method in [22]. It can be seen that the maximum lateral displacement and bending moment for the pile obtained using the three calculation methods were located near the tunnel axis, both the front pile and the back pile. The calculation results with and without considering the lateral soil were very similar to those from GEPAN when using the method presented in this paper to calculate the lateral displacement of adjacent group piles caused by shield tunnel excavation. What is more, the results in this work were more accurate than those from Zhang [22], especially in calculating the lateral displacement of the back pile. Moreover, the error in calculating the bending moment of the front pile was significantly smaller than in the work of Zhang [22]. Consequently, what has been discussed above proves the accuracy of the method proposed in this work.

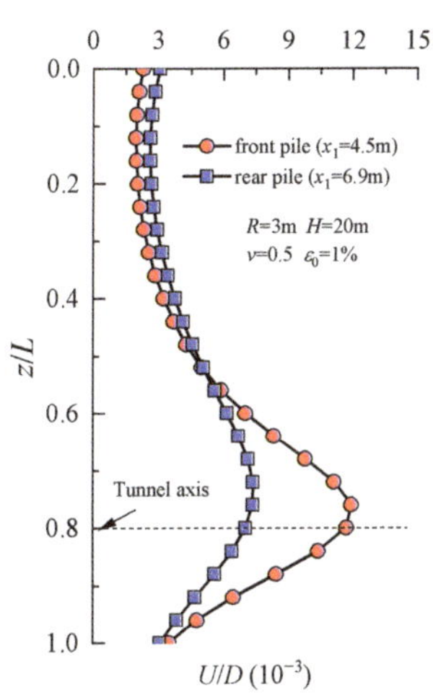

Figure 6. Lateral displacement of soil at the center position of the pile.

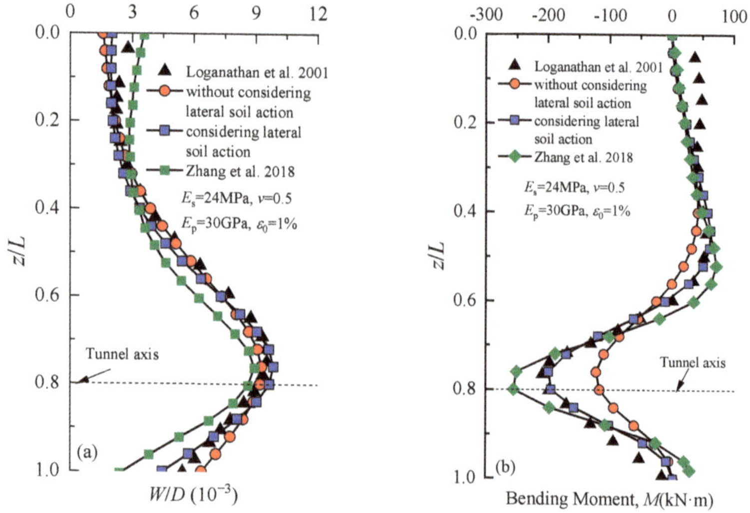

Figure 7. Comparison of results between the present method and others: (**a**) lateral displacement and (**b**) bending moment of the front pile [18,22].

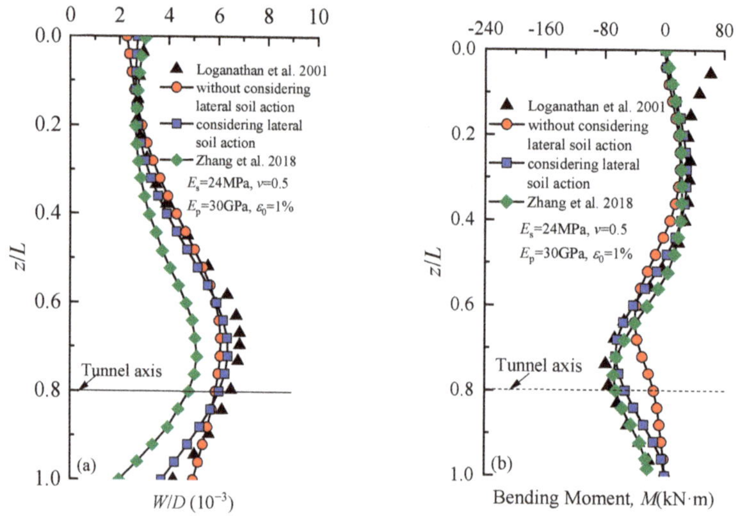

Figure 8. Comparison of results between the present method and others: (**a**) lateral displacement and (**b**) bending moment of the back pile [18,22].

5. Discussions

To analyze the impact of various factors on the deflection and bending moment of group piles using the method proposed in this paper, the following example was taken for analysis. The diameter R and the depth of the tunnel axis were 3 m and 20 m, respectively. A ground loss ratio ε_0 of 2.5% was considered. The soil's elastic modulus and Poisson ratio were 20 MPa and 0.3, respectively. The diameter D and the depth of the single pile were 0.6 m and 25 m, respectively. The elastic modulus of the single pile was selected as 30 GPa. Assuming that there were two piles with unconstrained top and tip, the front pile and back pile were located 5 m and 8 m from the tunnel axis, respectively. Moreover, the other parameters remained unchanged when analyzing a certain parameter's influence on the pile foundation's vertical displacement in the following.

5.1. Influence of the Stiffness of Soil Spring

Figure 9 shows the lateral displacement and bending moment of the front pile with different stiffnesses of soil spring *c*. This parameter represents the stiffness of the soil spring between the shear layer and the pile in the Kerr foundation model. It can be seen that the stiffness of soil spring *c* had a significant influence on the lateral displacement of the adjacent pile. Both the lateral displacement and bending moment of the front pile decreased with the increase in soil spring stiffness *c*, and the magnitude of the impact also gradually decreased. Increasing the soil spring stiffness resulted in a more significant reaction force for the Kerr foundation to the pile, reducing the front pile's displacement and bending moment.

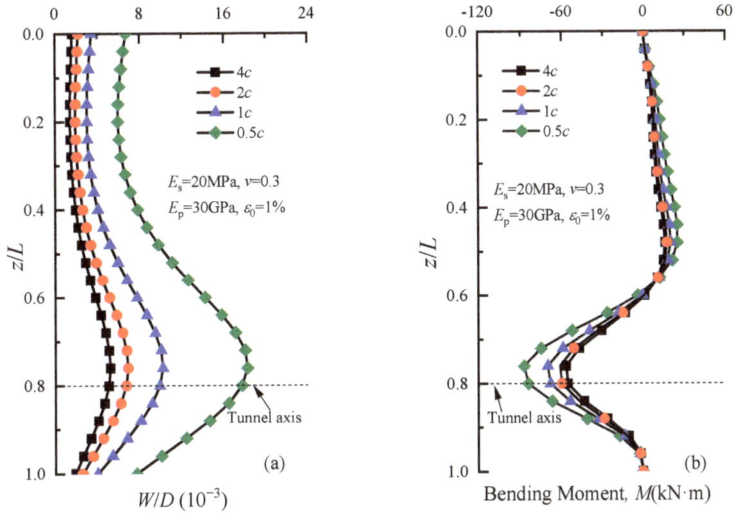

Figure 9. Variations of (**a**) lateral displacement and (**b**) bending moment of the front pile with different stiffnesses of soil spring.

5.2. Influence of Pile–Tunnel Distance

Figure 10 shows the front pile's lateral displacement and bending moment as the distance between the front pile and the tunnel axis were increased from 5 m to 8 m, while the pile spacing remained unchanged. The front pile's maximum lateral displacement and bending moment occurred at the centerline of the tunnel. As the distance between the front pile and tunnel increased, the front pile's maximum lateral displacement and bending moment decreased. This happened because the lateral displacement of the soil decreased, and the impact on the adjacent pile lessened.

5.3. Influence of the Ground Loss Ratio

Figure 11 illustrates the front pile's lateral displacement and bending moment under varying ground loss ratios ε_0. The graph in Figure 11 shows that, as the ground loss ratio increased from 0.5% to 2%, the front pile's lateral displacement and bending moment gradually increased. It is understandable that a higher ground loss ratio (ε_0) results in more significant lateral displacement of the soil, due to the excavation of the shield tunnel. This displacement put more load on the adjacent pile, resulting in greater lateral displacement and bending moment of the front pile.

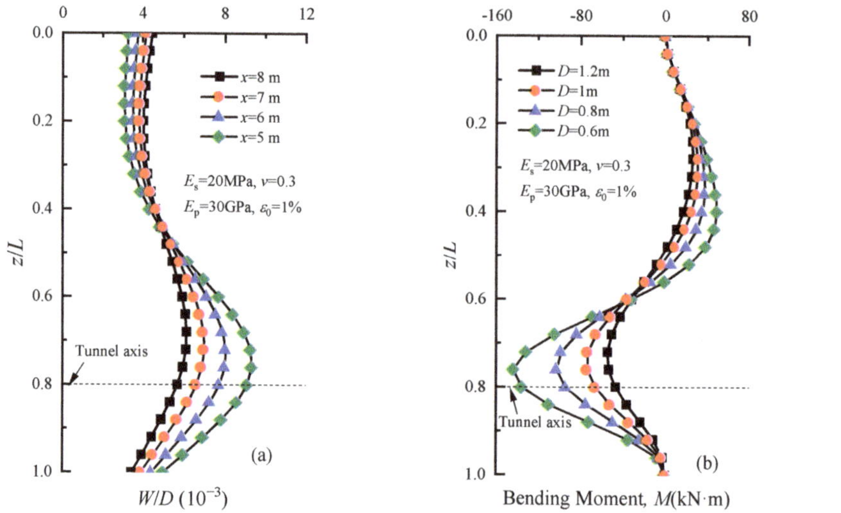

Figure 10. Variations of (**a**) lateral displacement and (**b**) bending moment of the front pile with different distances between the pile and tunnel.

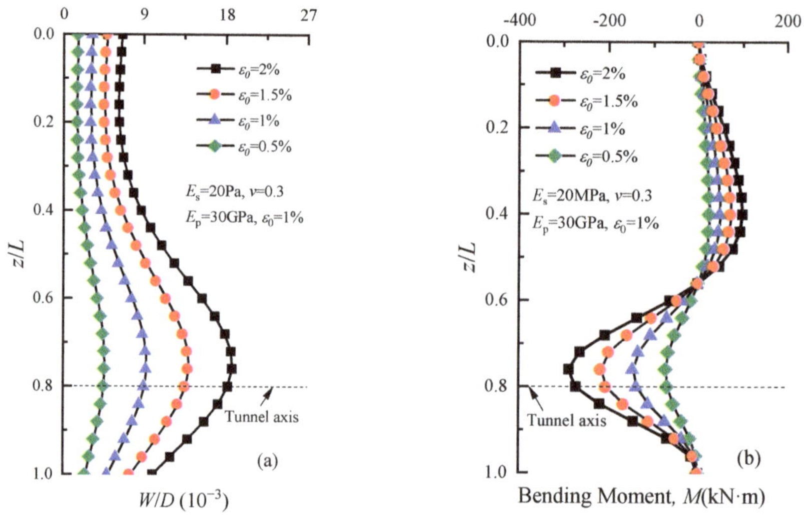

Figure 11. Variations in the (**a**) lateral displacement and (**b**) bending moment of the front pile with different ground losses.

5.4. Influence of Pile Diameter

The graph in Figure 12 illustrates how the front pile's lateral displacement and bending moment changed with different pile diameters. It is clear from the graph that changes in pile diameter had a significant impact on both lateral displacement and bending moments of the front pile. As the pile diameter D increased from 0.4 m to 1.0 m, the lateral displacement decreased slowly, while the bending moment increased. This is because the flexural rigidity EI increased with the diameter, leading to less deformation under the same load strength due to soil displacement but also to an increase in bending moment.

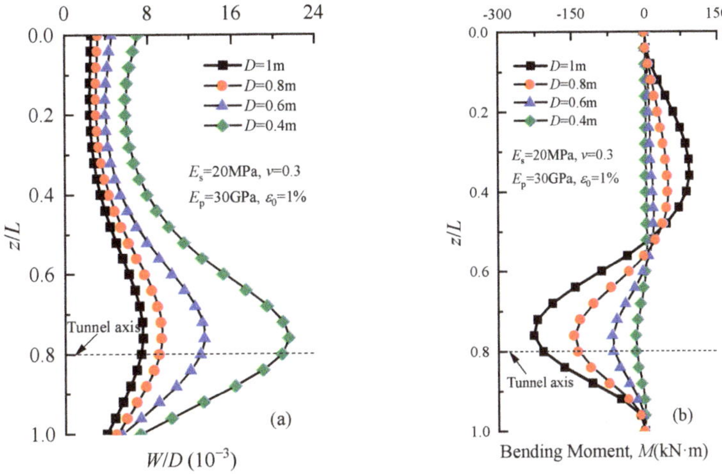

Figure 12. Variations in (**a**) lateral displacement and (**b**) bending moment of the front pile with different pile diameters.

6. Conclusions

This paper proposed a simplified method for simulating the pile–soil interaction caused by shield tunnel excavation. The process is based on the Kerr foundation model with three parameters and considers the influence of lateral soil displacements beside the pile. Theoretical solutions are presented to evaluate the effects of tunneling on the lateral displacements and bending moments of a single pile and group piles.

The presented method was validated by comparing it with existing solutions obtained from the boundary element program GEPAN. In addition, the proposed method was used to calculate the displacements and bending moments of a single pile and group piles induced by tunneling, while considering the effects of lateral soil displacements. A parametric analysis was conducted to assess the influence of various factors, including the stiffness of soil spring *c*, pile–tunnel distance, ground loss ratio ε_0, and pile diameter D on group piles' deflection and bending moment.

After an in-depth discussion of the results, the following conclusions can be drawn:

- For a more accurate calculation of the bending moment of a pile during shield tunnel excavation, it is essential to consider the influence of lateral soil displacements around the pile. Ignoring this influence can result in less accurate calculations. Therefore, accounting for the impact of lateral soil around the pile is crucial for achieving higher accuracy;
- The lateral displacement of the pile caused by tunnel excavation increases and then decreases with depth, with the maximum displacement and bending moment occurring at the depth of the tunnel axis;
- Increasing soil spring stiffness or pile diameter reduces lateral displacement but increases the bending moment;
- The pile's lateral displacement and bending moment decrease with increasing pile–tunnel distance, while they increase with an increase in the ground loss ratio.

Author Contributions: Conceptualization, N.W.; Software, H.J.; Formal analysis, H.J.; Writing—original draft, H.J.; Writing—review & editing, N.W., H.D. and L.G.; Funding acquisition, N.W. and H.D. All authors have read and agreed to the published version of the manuscript.

Funding: This research was funded by the National Natural Science Foundation of China, grant number 52168047 and 52208344.

Data Availability Statement: Not applicable.

Conflicts of Interest: The authors declare no conflict of interest.

References

1. Basile, F. Effects of tunnelling on pile foundations. *Soils Found.* **2014**, *54*, 280–295. [CrossRef]
2. Boonyarak, T.; Phisitkul, K.; Ng, C.W.; Teparaksa, W.; Aye, Z.Z. Observed ground and pile group responses due to tunneling in Bangkok stiff clay. *Can. Geotech. J.* **2014**, *51*, 479–495. [CrossRef]
3. Loganathan, N.; Poulos, H.; Stewart, D. Centrifuge model testing of tunnelling-induced ground and pile deformations. *Geotechnique* **2000**, *50*, 283–294. [CrossRef]
4. Marshall, A.M.; Mair, R.J. Tunneling beneath driven or jacked end-bearing piles in sand. *Can. Geotech. J.* **2011**, *48*, 1757–1771. [CrossRef]
5. Ng, C.W.W.; Lu, H.; Peng, S. Three-dimensional centrifuge modelling of the effects of twin tunnelling on an existing pile. *Tunn. Undergr. Space Technol.* **2013**, *35*, 189–199. [CrossRef]
6. Ng, C.W.W.; Soomro, M.A.; Hong, Y. Three-dimensional centrifuge modelling of pile group responses to side-by-side twin tunnelling. *Tunn. Undergr. Space Technol.* **2014**, *43*, 350–361. [CrossRef]
7. Lu, H.; Shi, J.; Ng, C.W.; Lv, Y. Three-dimensional centrifuge modeling of the influence of side-by-side twin tunneling on a piled raft. *Tunn. Undergr. Space Technol.* **2020**, *103*, 103486. [CrossRef]
8. Chiang, K.-H.; Lee, C.-J. Responses of single piles to tunneling-induced soil movements in sandy ground. *Can. Geotech. J.* **2007**, *44*, 1224–1241. [CrossRef]
9. Franza, A.; Marshall, A.M. Centrifuge and real-time hybrid testing of tunneling beneath piles and piled buildings. *J. Geotech. Geoenvironmental Eng.* **2019**, *145*, 04018110. [CrossRef]
10. Lee, G.T.; Ng, C.W. Effects of advancing open face tunneling on an existing loaded pile. *J. Geotech. Geoenvironmental Eng.* **2005**, *131*, 193–201. [CrossRef]
11. Liyanapathirana, D.; Nishanthan, R. Influence of deep excavation induced ground movements on adjacent piles. *Tunn. Undergr. Space Technol.* **2016**, *52*, 168–181. [CrossRef]
12. Soomro, M.A.; Hong, Y.; Ng, C.W.W.; Lu, H.; Peng, S. Load transfer mechanism in pile group due to single tunnel advancement in stiff clay. *Tunn. Undergr. Space Technol.* **2015**, *45*, 63–72. [CrossRef]
13. Jongpradist, P.; Kaewsri, T.; Sawatparnich, A.; Suwansawat, S.; Youwai, S.; Kongkitkul, W.; Sunitsakul, J. Development of tunneling influence zones for adjacent pile foundations by numerical analyses. *Tunn. Undergr. Space Technol.* **2013**, *34*, 96–109. [CrossRef]
14. Hong, Y.; Soomro, M.A.; Ng, C.W.W. Settlement and load transfer mechanism of pile group due to side-by-side twin tunnelling. *Comput. Geotech.* **2015**, *64*, 105–119. [CrossRef]
15. Fu, J.; Yang, J.; Zhu, S.; Shi, Y. Performance of jet-grouted partition walls in mitigating the effects of shield-tunnel construction on adjacent piled structures. *J. Perform. Constr. Facil.* **2017**, *31*, 04016096. [CrossRef]
16. Chen, L.T.; Poulos, H.; Loganathan, N. Pile responses caused by tunneling. *J. Geotech. Geoenvironmental Eng.* **1999**, *125*, 207–215. [CrossRef]
17. Loganathan, N. Analytical Prediction for Tunneling-Induced Ground Movements in Clays. *J. Geotech. Geoenvironmental Eng.* **1998**, *124*, 846–856. [CrossRef]
18. Loganathan, N.; Poulos, H.; Xu, K. Ground and pile-group responses due to tunnelling. *Soils Found.* **2001**, *41*, 57–67. [CrossRef]
19. Huang, M.; Zhang, C.; Li, Z. A simplified analysis method for the influence of tunneling on grouped piles. *Tunn. Undergr. Space Technol.* **2009**, *24*, 410–422. [CrossRef]
20. Mu, L.; Huang, M.; Finno, R.J. Tunnelling effects on lateral behavior of pile rafts in layered soil. *Tunn. Undergr. Space Technol.* **2012**, *28*, 192–201. [CrossRef]
21. Pasternak, P. On a new method of an elastic foundation by means of two foundation constants. *Gos. Izd. Lit. Po Stroit. I Arkh.* **1954**, 1–56.
22. Zhang, Z.; Huang, M.; Xu, C.; Jiang, Y.; Wang, W. Simplified solution for tunnel-soil-pile interaction in Pasternak's foundation model. *Tunn. Undergr. Space Technol.* **2018**, *78*, 146–158. [CrossRef]
23. Kerr, A.D. A study of a new foundation model. *Acta Mech.* **1965**, *1*, 135–147. [CrossRef]
24. Zhang, D.-M.; Huang, Z.-K.; Li, Z.-L.; Zong, X.; Zhang, D.-M. Analytical solution for the response of an existing tunnel to a new tunnel excavation underneath. *Comput. Geotech.* **2019**, *108*, 197–211. [CrossRef]
25. Vesić, A.B. Bending of beams resting on isotropic elastic solid. *J. Eng. Mech. Div.* **1961**, *87*, 35–53. [CrossRef]
26. Tanahashi, H. Formulas for an Infinitely Long Bernoulli-Euler Beam on the Pasternak Model. *Soils Found.* **2004**, *44*, 109–118. [CrossRef]
27. Randolph, M.; Wroth, C. Analysis of Deformation of Vertically Loaded Piles. *J. Geotech. Geoenvironmental Eng.* **1978**, *104*, 1465–1488. [CrossRef]

buildings

MDPI

Article

A Seismic Risk Assessment of Concrete-Filled Double-Skin Steel Tube (CFDST) Frames with a Beam-Only-Connection for Reinforced Concrete Shear Walls (BRWs)

Hongyu Sun [1], Yi Hu [1,*], Junhai Zhao [2], Da Wang [1] and Yi Liu [1]

[1] School of Civil Engineering, Central South University of Forestry and Technology, Changsha 410004, China; 20211200551@csuft.edu.cn (H.S.); yxwang2006@yeah.net (D.W.); 20221200567@csuft.edu.cn (Y.L.)
[2] School of Civil Engineering, Chang'an University, Xi'an 710061, China; zhaojh@chd.edu.cn
* Correspondence: hyi_1991@163.com

Abstract: The beam-only connected reinforced concrete shear wall (BRW) is used as a reinforcing component to enhance the seismic performance of concrete-filled, double-skin steel tube (CFDST) frames. The effects of the BRW on seismic risks of CFDST frames are investigated. Three performance levels of limit states are defined and described according to the typical failure of test specimens and the existing definition of current guidance. A simplified numerical model is calibrated for CFDST frame-BRW structures, and test results validate it. Nonlinear dynamic analyses on a nine-story CFDST-BRW building are performed to investigate the effects of BRW on reducing the seismic risk of CFDST buildings. The results show that the BRW reduces the probability of collapse of the CFDST frame to 2.76% in 50 years, which can effectively reduce the probability of different degrees of damage in the service cycle of the structure. The results provide information for risk-informed decision-making on the design of CFDST frame-BRW structures.

Keywords: concrete-filled double-skin steel tube; assembled beam-only connected reinforced concrete shear walls; seismic risk reduction; performance-based seismic design

Citation: Sun, H.; Hu, Y.; Zhao, J.; Wang, D.; Liu, Y. A Seismic Risk Assessment of Concrete-Filled Double-Skin Steel Tube (CFDST) Frames with a Beam-Only-Connection for Reinforced Concrete Shear Walls (BRWs). *Buildings* **2023**, *13*, 2378. https://doi.org/10.3390/buildings13092378

Academic Editor: Binsheng (Ben) Zhang

Received: 28 August 2023
Revised: 11 September 2023
Accepted: 12 September 2023
Published: 19 September 2023

1. Introduction

The traditional concrete-filled steel tubular (CFST) frame structures rely on the connection of CFST frame joints (finger beam-column joint) to achieve lateral resistance. Once the CFST structure joints are damaged, they will collapse. Therefore, the frame-shear wall structure system and seismic design method have gradually attracted attention. Many researchers have conducted experimental, theoretical and numerical studies on the design methods of different structures.

So far, many experimental and theoretical studies have been carried out on the performance enhancement of CFST columns and frames. Wang [1] designed thin steel circular CFSTs. The specimens were tested, and their seismic performances were evaluated. Skalomenos [2] proposed using the HFD method to guide the seismic assessment of CFST frames. Gan [3] studied the mechanical properties of shear connectors in CFST frame structures. Lai [4] and Wang [5] used high-strength materials to improve the seismic capacities of CFST frames. Bai [6] established a model of a CFST frame and verified the correctness of the connection parts of the numerical model through experiments. Ren [7] and Jia [8] studied the typical working relationships between energy dissipation devices, buckling restrained braces, and CFST frames.

However, because the frame structure realizes the lateral resistance through the connections of beam and column members, only one seismic defense is included, and the scope of the application is limited. People gradually advocate using shear walls to improve the performance of frame structures.

For this reason, many shear wall types, frame connection methods, and seismic theories have been proposed. Nie [9] suggested using a composite shear wall to strengthen a CFST frame. The cooperative working relationship between the composite shear wall and the CFST frame was deduced. Wang [10] conducted seismic performance tests on concrete-filled steel tubular infilled walls. Guo [11] proposed a connection method between the CFST frames and partially infilled walls and conducted experimental research on their seismic performance. Hu [12] studied the cyclic performance of a CFDST frame strengthened by a BSW. Bian [13] proposed an assembled structure composed of a new frame-shear wall structure. Hu [14] proposed a fully precast CFSDT frame, which is only connected to the BSW system. An SPSW with a beam-only connection for this system was presented. Xu [15] proposed a precast CFST frame support system by using reinforced concrete beams and a shaking table test. Skalomenos [16] studied different modeling methods concerning the seismic vulnerability of CFST integral structures with or without rigid PZs. The influence of varying modeling methods on the structural vulnerability was quantified. Leon [17] proposed the concept of resilience of CFST composite structures and suggested that the resilience design of systems under different levels of earthquakes can be realized by determining the performance level indicators. Kamaris [18] provided a simplified seismic design method for CFST structures. Ahmadi [19] established a vulnerability model of CFST structures and calculated the failure probability of the structure under earthquake actions, which was used as the risk data of structural performance-based design. Hu [20] assessed the seismic risk of the CFDST frames. Six typical extreme performance levels of the CFDST structure were determined. The models for the specimens were established. The seismic responses of these models to seismic excitation were studied by nonlinear time history analysis, and the limit capacity were determined by incremental dynamic analysis.

Although many different types of shear walls have been proposed, most of these shear walls are connected to the frames with complete filling. The additional stresses by the shear walls will directly act on the column, which are unfavorable to the seismic resistance of the column. In addition, although there is already research foundations for CFST frame-shear walls, there are almost no research reports on composite CFDST frame-shear walls. Based on the above existing research, we propose a CFDST frame-BRW structure.

This study aims to define and describe three types of performance limit states applicable to this type of structure through the typical failure characteristics of frame-BRW structures. Based on the test-validating method, the numerical analysis models of specific composite frame-BRW structures are established. The regional representative seismic waves are selected, and the nonlinear dynamic analysis of the nine-story structure is carried out. Then, based on the theory of seismic risk assessment, a structural probability calculation model is constructed to evaluate the seismic risk of the structure quantitatively. The results will provide ideas and risk data for the performance-based seismic design of assembled CFDST frame-BRW structures.

2. Cyclic Tests and Performance Limits

A total of two CFDST-BRW specimens were designed and tested. The BRW is the reinforced concrete shear wall. There were two test specimens investigated in this test.

These specimens were scaled down to half and were fabricated using the prefabrication method proposed in Ref. [20]. Details of these specimens are shown in Figure 1.

In the elastic stage, the CFDST frame and the BRW did not show obvious deformation. The state changed when the story drift ratio (SDR) was close to 1.0%, and the experimental curves showed an inflection point. Then, test specimens entered the elastic–plastic state. The cracks occurred on the BRW at this stage, resulting in the nonlinear behaviors of the test specimens. The specimens were significantly deformed as the SDR reached 3.0% and entered the failure stage. The loading capacity of all CFDST frame-BRW specimens began to degrade significantly, and the BRW gradually failed. The weld at the CFDST joint was torn, the obvious buckling deformation occurred on the stiffener of the CFDST joint and the plate at the beam flange, and the plate at the bottom of the column sustained local buckling. When the SDR reached 6.0%, the ultimate loading capacity of all specimens was obviously degraded, and all of the BRW failed. The CFTSD joints and column bottoms in the CFDST frame form obvious plastic hinges, and the structure was about to lose its bending resistance and cause overall deformation. The envelop curves of the specimens are shown in Figure 2.

(**a**) BF-BRW-1 (**b**) BF-BRW-2

Figure 1. Test specimens.

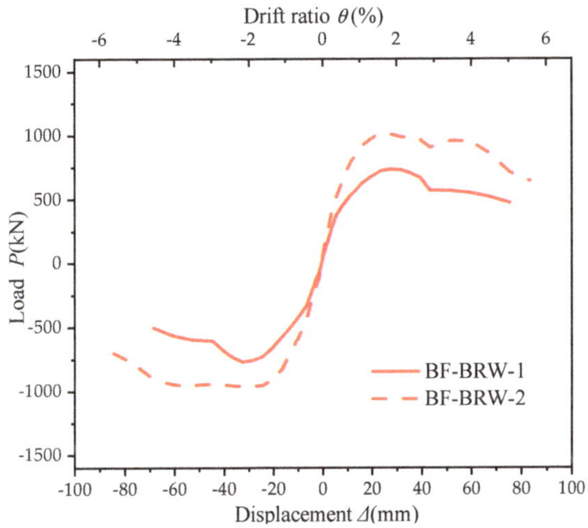

Figure 2. Envelop curves of the test specimens.

There are three typical mechanical phases.

(1) Stage I: from 0 to 1% drift. The CFDST frame-BRW structure can be approximately considered to be in the elastic stage. Although the BRW has small cracks, the deformation has little effect on their structural performance.

(2) Stage II: from 1 to 3% drift. Many obvious cracks appeared on the BRW, and the bolt connection between the BRW and CFDST frame was damaged to a certain extent. Local buckling of column and beam flanges of the CFDST frames was obvious, and the weld connections and flange plates near the high-strength bolts were torn.

(3) Stage III: from 3 to 6% drift. The BRW was obviously deformed, and it had basically separated from the CFDST frame and lost its loading capacity. The steel beam plate and high-strength bolts near the CFDST column were torn, and the CFDST frame joints and the bottom of the column formed a plastic hinge, which has basically lost its loading capacity.

The American design codes FEMA 356 [21], HAZUS [22], and SEAOC [23] have not defined the values of the limit state of the CFDST frame-BRW structure, and other design codes have not provided relevant reference information. We used the maximum inter-story drift ratio (ISDAmax) as the performance index to describe the limit state of the building structure. Combined with the definition of the CFDST frame defined by Ref. [20], and the meaning of the performance limits of the CFDST structure in FEMA 356 performance limit of the CFDST frame-BRW structure is defined as follows.

(1) Immediate Occupancy (IO): the structure needs to be used immediately after the earthquake. This stage is described for the CFDST frame-BRW structure, i.e., the BRW has small cracks, the steel beam flange or web exhibits a slight yield or buckling phenomenon, and no obvious fracture phenomenon is observed. The performance limit is defined as 1.0%.
(2) Structural Damage (SD): the structural or nonstructural components may be seriously damaged or partially collapsed, but the building still has a certain ability to resist collapse, which can ensure that the structure will not collapse. For the CFDST frame-BRW structure, this stage is described as follows. The BRW has many cracks, and the concrete at the bolt connection between the BRW and CFDST frame exhibits spalling phenomena. There are many steel plates yielding, buckling, or fracturing on the steel beam near the column, but they do not completely fail, and some members have obvious local buckling or plastic hinges. The performance limit is defined as 3.0%.
(3) Structural Damage (SD): the structure collapses due to its inability to withstand its own gravity or insufficient lateral stiffness. For the CFDST frame-BRW structure, this stage is described as follows. The BRW has basically separated from the CFDST frame and has lost its loading capacity. The steel beam plate in the CFDST frame is torn, the joint concrete is crushed, and the outer steel tube is severely buckled. The performance limit is defined as 6.0%.

3. Numerical Model and Validation

3.1. CFDST Framework Numerical Model and Verification

Hu established the test-validated numerical model of the CFDST framework [20]. Based on this model, a nine-story CFDST frame-BRW was constructed, and the influence of a BRW on the seismic risk of a nine-story CFDST frame was studied. A seismic risk assessment was carried out. To ensure the integrity of the paper, the modeling and verification of the CFDST frame structure are still explained.

3.1.1. CFDT Column, Steel Beam, and CFDST Joints

The CFDST frame structure model is established by using the OpenSees. Nonlinear elements are used to simulate the performances of CFDST columns and beams, and the CFDST column filled with concrete uses the Concrete 02 material, while the steel tube and beam use the Steel 02 material. The Kent–Scott–Park damage model is used for concrete materials to simulate the hysteretic characteristics and material degradation properties accurately.

Three nonlinear spring elements, i.e., a horizontal $K_{H,}$ a vertical K_V, and a rotating K_θ, are used to simulate the CFDST joint model in Figure 3, and the mechanical behaviors of the nonlinear springs used to simulate the CFDST joints and nonlinear springs are simulated by the Zerolength element in OpenSees. In the numerical model, the model of the horizontal spring corresponds to two nodes at the connection position of the beam and column, and the modeling is achieved by maintaining the same horizontal displacement of the two nodes.

According to the failure modes observed during the test, the local buckling and fracture behavior of the steel beams occurred, resulting in strength and stiffness degradations of CFDST joints. Therefore, a simple nonlinear model cannot accurately simulate the mechanical behaviors of steel beams. An energy-based improved IK steel damage model developed by Lignos [24] was selected to simulate the intricate mechanical behaviors of steel beams accurately. The model is based on the Ibarra–Krawinkler (IK) [24] and more than 300 wide-flange steel beam tests.

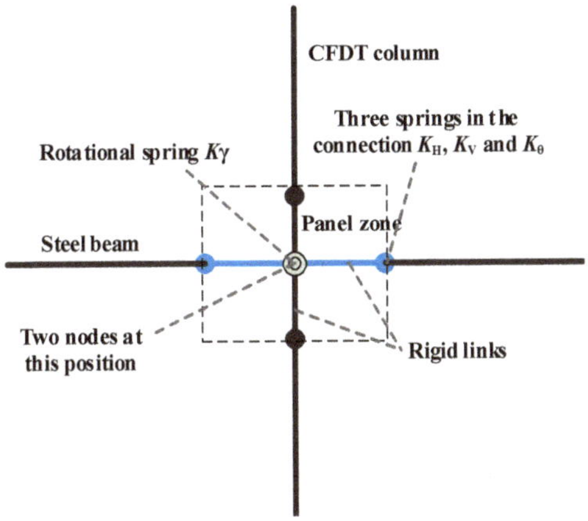

Figure 3. The PZ model of CFDST frame beam-to-column intermediate joints.

The monotonic curve model of the improved IK model was adopted, as shown in Figure 4. The curve contains the yield moment (M_y), capping moment (M_c), residual moment (M_r), yield rotation (θ_y), pre-capping plastic rotation (θ_p), post-capping plastic rotation (θ_{pc}) and ultimate rotation capacity (θ_u). The strength and nonlinear deformation values should be appropriately reduced since the specimen is subjected to a reciprocating load. The values that need to be modified are shown as follows: (1) M_c is defined as 0.9 times the initial value but not less than M_y; (2) θ_p is determined to be 0.7 times the initial value; (3) θ_{pc} is determined to be 0.5 times the initial value.

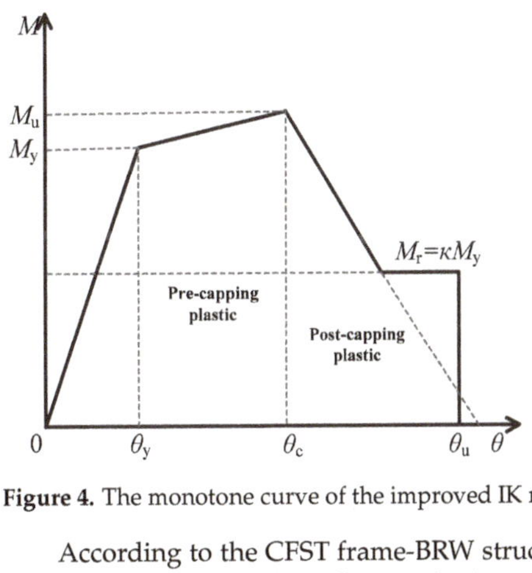

Figure 4. The monotone curve of the improved IK model.

According to the CFST frame-BRW structure test, The main failure behavior of the beam is located near the column. An improved IK model simulates the degradation behavior of steel beams by using a rotating spring. The vertical spring is modeled as being fully elastic. Therefore, the steel beam is transformed into a bilinear strengthening model.

3.1.2. Panel Zone

This paper establishes the panel zone (PZ) model to simulate the nonlinear behaviors of the CFDST joint panel. The PZ model should consider the influence of multiple factors, such as concrete cracking, concrete crushing, and steel yield, and the model should be based on the existing members of OpenSees.

The three-linear shear deformation model established by Skalomenos [16] modeled the PZ in the CFT column connection. In this idealized model, the total shear strength is the sum of the strengths of the steel tube and the concrete in the panel under the same deformation. To be sure of the shear yield strength and post-yield stiffness of the ideal model, this study considers the performances of the joint zone of the CFT column-wide flange beam bolted bending moment connection and the inner end plate and the CFT joint of the through-beam [25]. The PZ model of CFDST column joints is established, as shown in Figure 3. A rotating spring between the beam–column represents the relative rotation. The rotating spring is modeled by the Zerolength element. Two stiffness bar elements are used to simulate the relative rotation between the two motions of beams and columns and are modeled by stiffness-link elements. The rotational spring represents the mechanical behaviors of the joint, including rotational stiffness and shear capacity. Therefore, it is essential to model the stiffness and shear capacity of the spring and then simulate the nonlinear behaviors of the PZ. By determining the stiffness and shear loading capacity of the spring, the complex force relationship of the joint can be simplified into a functional relationship of multiple line segments. Then, it can be simplified into a relatively simple mathematical calculation problem.

The horizontal shear capacity of the PZ includes the shear contributions provided by steel tube and concrete. The shear capacity Vu is the sum of the shear capacities of steel tube and concrete:

$$V_u = V_s + V_c = A_{sv}(\frac{f_y}{\sqrt{3}}) + 1.99\sqrt{f_c'}A_{cv} \tag{1}$$

where A_{sv} and A_{cv} are the effective shear extents of the steel tube and concrete, respectively, and f_y and f_c' are the steel tube yield strength and the core concrete compressive strength, respectively.

The core concrete is in a three-dimensional compressive stress state under the action of gravity load. Figure 5 is the mechanical model of core concrete without considering the internal and external steel pipes under gravity load.

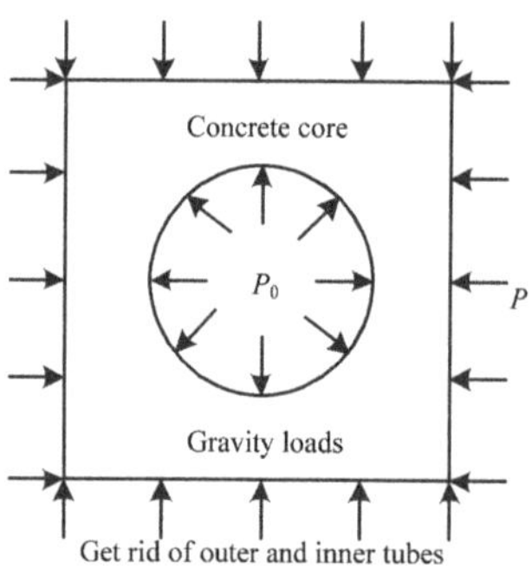

Figure 5. The stress state of concrete filled in the CFDST column.

The functional relationship between the compressive strength of core concrete and the standard strength of concrete is as follows:

$$f_c' = f_c + k_c(P + \frac{2\zeta P_0 B}{D\sqrt{\pi}}) \tag{2}$$

$$\zeta = 66.474(\frac{t}{B})^2 + 0.992\frac{t}{B} + 0.416 \tag{3}$$

where f_c is the standard compressive strength of concrete, k_c is the strength improvement coefficient of concrete under lateral restraint, and ζ is the constraint reduction coefficient, which can be calculated by Equation (3). P and P_0 are the pressures caused by the outer and inner steel tubes, respectively. B and t are the width and thickness of the outer steel tube, respectively, and D is the diameter of the inner steel tube.

Figure 6 shows the idealized trilinear model of the PZ of CFDST frame joints. The yield point of the PZ is the yield point of the steel tube.

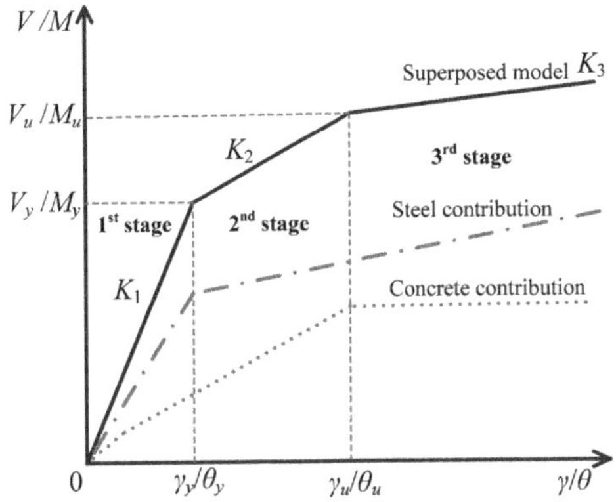

Figure 6. Trilinear model of the CFDST frame beam-column PZ.

The calculation of the shear deformation is shown as follows:

$$\begin{gathered}\gamma_y = \kappa_{s1} \times (\frac{V_{s1}}{A_{sv1} \times G_s}) = \kappa_{s2} \times (\frac{V_{s2}}{A_{sv2} \times G_s})\partial y \\ V_{s1} + V_{s2} = V_s\end{gathered} \tag{4}$$

where κ_{s1} and κ_{s2} are the shear coefficients of the square and circular steel tubes, which are equal to 1.2 and 10/9, respectively. A_{sv1} and A_{sv2} are the shear extents of the outer and inner tubes, respectively. G_s is the shear modulus of the steel tube. When the shear force of the PZ reaches 60% of the ultimate loading capacity V_u, the PZ yields. The inelastic stiffness of the second stage K_2 is 20% of the initial stiffness of the first stage K_1. The bending moment–rotation (M-θ) relationship of the rotating spring can be converted from the shear-deformation relationship mentioned above.

The bending moment M and the rotation angle θ can be calculated by the following formulas as:

$$M = V \times B,\ K_\theta = K_\gamma \times B,\ \theta = \gamma \tag{5}$$

3.2. CFDST Frame-BRW Structure Numerical Model and Verification

In the CFDST frame-BRW structure, the BRW and CFDST frame are connected by high-strength bolts. Therefore, many nonlinear behaviors are involved in seismic analysis. Considering the use of classical finite element software ANSYS or ABAQUS to finely model the CFDST frame-BRW structure, it will not only cause the numerical model to be complicated and the calculation cost to be too high, but also may cause the numerical model to be difficult to converge due to various nonlinearities. The simplified analysis method is used to establish the numerical analysis model. The simplified method utilized in Ref. [20] is used to model the CFDST framework. We recommend using two cross-arranged nonlinear springs to consider the nonlinear behaviors of the BRW. The seismic performance simulation analysis of the CFDST frame-BRW structure is still based on the OpenSees.

The two-node link element is used to simulate the nonlinear springs, and the hysteresis parameters of the BRW are input through the Pinching04 material model. The length of the two-node link element can be zero or nonzero. The element can simulate the deformation of the structure with 1 to 6 degrees of freedom. The two-node link element used in this paper only considers the axial deformation, that is, the contribution of the BRW to the lateral stiffness of the CFDST frame-BRW structure is simulated by the axial deformation. Pinching04 is a uniaxial material model defined by 39 parameters in Figure 7.

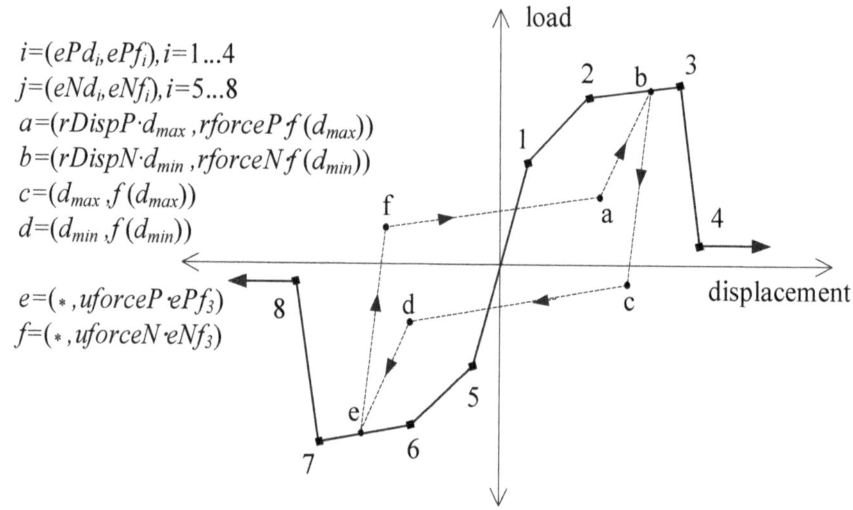

Figure 7. Hysteresis model of Pinching04 material.

The element can characterize the deformation response of the structural member under uniaxial tension-compression loads. They can also represent the loading capacity and stiffness degradation behavior of the member under cyclic loading. The Pinching04 material model uses 16 parameters to define the skeleton curve of the members (ePf1, ePd1, ePf2, ePd2, ePf3, ePd3, ePf4, ePd4, eNf1, eNd1, eNf2, eNd2, eNf3, eNd3, eNf4 and eNd4) and uses 6 parameters to define the hysteretic characteristics of the members under a reciprocating load (uForceP, uForceN, rDispP, rDispN, rForceP and rForceN). The loading capacity degradation behaviors of the member under cyclic loading is defined by five parameters (gF1, gF2, gF3, gF4, and gF5). The unloading stiffness degradation of the member is controlled by five parameters (gK1, gK2, gK3, gK4, and gKLim). The loading stiffness degradation of the member is controlled by five parameters (gD1, gD2, gD3, gD4, and gDLim). The maximum degradation degree of each cycle is limited by two parameters (gE and dmgType). In addition, the Pinching04 material model can consider positive and negative loadings, so the Bauschinger effect of the model can be considered. For each loading direction, the skeleton curve can be defined by four parameters. Similarly, each direction can also define the hysteretic parameters, loading capacity, and stiffness degradation parameters. Because the Pinching04 material model contains many parameters, it can simulate the hysteretic behavior of complex structures under an earthquake.

Figure 8 is a geometric transformation relationship between the load–displacement curve of the element and the load–displacement curve of the BRW, including the force and displacement borne by the element:

$$F' = F/2\cos\theta \tag{6}$$

$$\Delta_w' = \Delta_w \cdot \cos\theta \tag{7}$$

where F and F' are the horizontal forces borne by the wall and the axial force borne by a two-node link element, respectively; Δ_w and $\Delta_w{}'$ are the horizontal displacements of the BRW and the axial deformation of a two-node link element, respectively; θ is the angle between the two-node link element and the steel beam and represents the angle between the force and deformation direction of the two-node link and the horizontal direction.

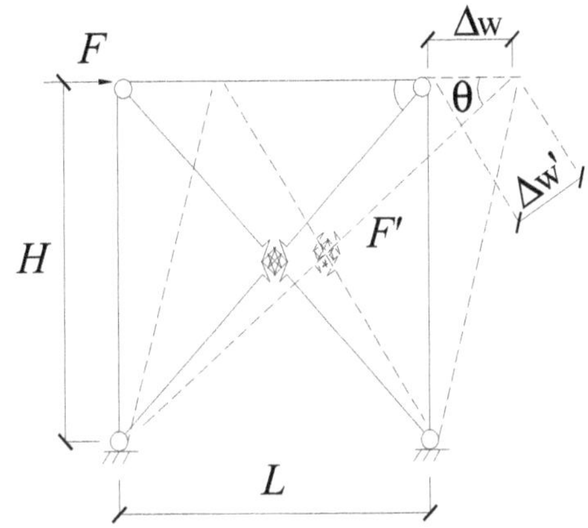

Figure 8. Relationship between the force and deformation.

After selecting the element and material model of the simulated BRW, it is necessary to ensure the parameters in the material model, in which the skeleton curve is related to the loading capacity and deformation capacity of the BRW. To match the Pinching04 material model, this paper recommends a fivefold line model for the skeleton curve of the BRW connected to the beam end. There are four key points and five stages in Figure 9. The stage is known as the elastic stage, in which the BRW concrete does not crack, and the high-strength bolted connection between the BRW and the CFDST frame is not torn.

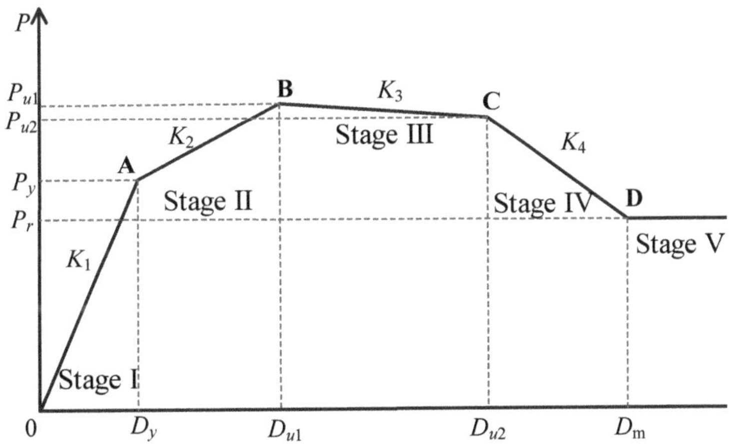

Figure 9. Definition of the skeleton curve for the BRW.

The loading capacity P_y and the displacement D_y at point A are defined as the loading capacity corresponding to the cracking of concrete in the BRW, and the displacement is the ratio of the loading capacity to the elastic lateral stiffness of the BRW. Because the test model involves the initial defects of the BRW, the uncertainty of the material properties, and the bolt connection between the BRW and the CFDST frame, the ideal state theory cannot sufficiently reflect the stress state of the CFDST frame-BRW structure. To this end, this paper draws on the simplified processing method adopted by the energy-based improved IK damage model proposed by Lignos [26]. It uses the test data to determine the skeleton curve's loading capacity and deformation parameters, represented by a simple constant coefficient. According to the analysis of the results of the two test models, the following key points are defined.

(1) Point A is defined as follows, i.e., the loading capacity P_y of the BRW is 0.59 times the peak loading capacity P_{u1} of the BRW. The SDR is 0.6%, that is, the displacement is 9.0 mm.
(2) Point B is defined as follows, i.e., the peak loading capacity P_{u1} of the BRW is calculated according to the simplified calculation method. The SDR of the BRW is 1.3%, and the displacement is 19.5 mm.
(3) Point C is defined as follows, i.e., the loading capacity P_{u2} of the BRW is 0.75 times (BF-BRW-1) and 0.85 times (BF-BRW-2) of the peak loading capacity P_{u1} of the BRW, and the SDRs are 2.6% (BF-BRW-1) and 3.3% (BF-BRW-2), that is, the displacements are 39.0 mm (BF-BRW-1) and 49.5 mm (BF-BRW-2), respectively.
(4) Point D is defined as follows, i.e., the loading capacity P_m of the BRW is 0.27 times that of the peak loading capacity P_{u1} of the BRW. The SDR is 5.7%, that is, the displacement is 85.5 mm.

When the skeleton curve parameters of the BRW are determined, the skeleton curve parameters of the BRW are converted into the skeleton curve parameters of the two-node link element according to Equations (8) and (9) and are input into the simplified model of the structure to realize the simulation analysis of the test model. Among them, ePf1, ePf2, ePf3, and ePf4 correspond to the forces at points A, B, C, and D after the geometric relationship is transformed; ePd1, ePd2, ePd3, and ePd4 correspond to the displacements at points A, B, C and D after geometric transformation, respectively.

$$\lambda_j = \frac{P^i_{\max}}{P^1_{\max}} \tag{8}$$

$$K_i = \frac{\sum\limits_{j=1}^{2} P^i_j}{\sum\limits_{j=1}^{2} \Delta^i_j} \tag{9}$$

where $P^i_{\max}$ represents the peak load of the specimen under the ith reciprocating load; $P^1_{\max}$ denotes the peak load of the specimen in the 1st loading stage; K_i is the secant stiffness of the test specimen in the ith loading stage; P^i_j indicates the peak load capacity of the test specimen in the jth direction (forward loading and reverse loading) when the ith level is loaded; and Δ^i_j indicates the displacement corresponding to the peak loading capacity of the specimen in the j direction when the ith stage is loaded.

In addition to the forces and displacements at points A, B, C, and D, the hysteretic parameters, loading capacity, and stiffness degradation parameters of the two-node link element need to be determined. Analysis of the hysteretic curves measured in the test, the hysteretic parameters, loading capacity, and stiffness degradation parameters of the two-node link element can be obtained as follows.

For the BRW, the hysteretic parameters are uForceP = 0.2, uForceN = 0.2, rDispP = 0.8, rDispN = 0.8, rForceP = 0.95 and rForceN = 0.95, the loading capacity degradation parameters are gF1 = 0.0, gF2 = 0.0, gF3 = 0.0, gF4 = 0.0 and gF5 = 0.0, and the unloading stiffness degradation parameters are gK1 = 0.0, gK2 = 0.0, gK3 = 0.0, gK4 = 0.0 and gKLim = 0.0. The loading stiffness degradation parameters gD1 = 0.0, gD2 = 0.0, gD3 = 0.0, gD4 = 0.0 and gDLim = 0.0, the maximum degradation gE = 10.0 and dmgType = Energy.

The above parameters are input into the numerical analysis model, and the CFDST frame-BRW structure is numerically simulated and analyzed. The two CFDST frame-BRW structure tests completed are used to verify the numerical model as shown in Figure 10.

Since the CFDST frame connection and the BRW are represented by multiline segment functions, the predicted hysteresis curves show the prominent line segment shape characteristics. The simplified numerical model not only simulates the ultimate loading capacity of the test model but also has high accuracy for the degradation behaviors of the stiffness and loading capacity of the test model under cyclic loading. This shows that the model can accurately predict the seismic performances of the CFDST frame-BRW structure. In addition, compared with traditional finite element analysis software, such as ANSYS and ABAQUS, the calculation time of a single model using the simplified analysis method is less than 5 min, the calculation efficiency is high, and the convergence is good. Therefore, this method can be considered as an effective numerical method for seismic analysis and dynamic calculation of CFDST frame-BRW structures.

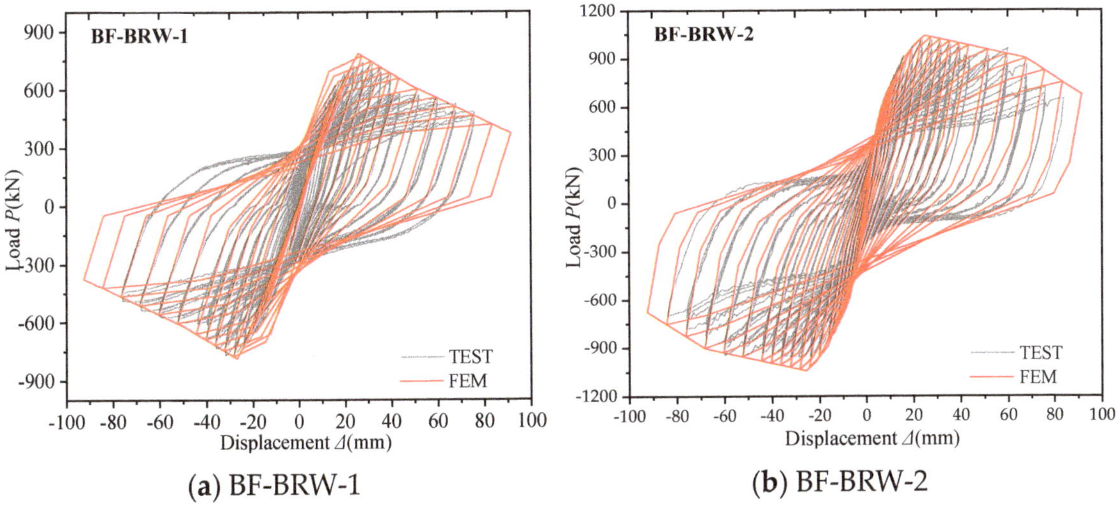

Figure 10. Comparison of FEM and test results.

4. CFDST Frame-BRW for Seismic Risk Assessment

The earthquake risk assessment theory includes the following step: (1) collect ground motion data; (2) dynamically analyze the target building and obtain its response; (3) measure the structural performance and determine the degree of damage; (4) establish a vulnerability curve; and (5) calculate the seismic risk and make design decisions.

According to Ref. [20], the probability of the limit state in the seismic risk assessment generated the following formula:

$$P[LS_i] \approx (k_0 m_R{}^{-k}) \exp[(k\beta_R)^2/2] \quad (10)$$

where $P[LS_i]$ is the occurring probability of each limit state in seismic evaluation; LS_i is the ith performance limit state of the structure; k and k_0 are the shape parameter and proportional parameter of the seismic hazard curve, respectively; and m_R β_R are the median and logarithmic standard deviation of seismic demand, respectively.

5. Earthquake Ground Motions

The seismic load and structural resistance of existing buildings are uncertain. The uncertainty of the seismic demand is dominant in the overall response. Therefore, The parameters of the numerical model are constants such as the yield stress and modulus. As China's earthquake disaster map and regional representative seismic wave selection are still in the research and development stage, this paper takes the Los Angeles area of the US as an example and selects 40 ground motions. These ground motions are representative ground motions set by the SAC project [22] for the Los Angeles (LA) area for seismic risk analysis of steel building structures on site. These records represent two risk levels: the excess probability within 50 years is 10% (la01-20) and 2% (la21-40), each group containing

20 ground motions. The response spectrum of ground motions la01 to la40 in Figure 11 represents two seismic risk levels in the Los Angeles area.

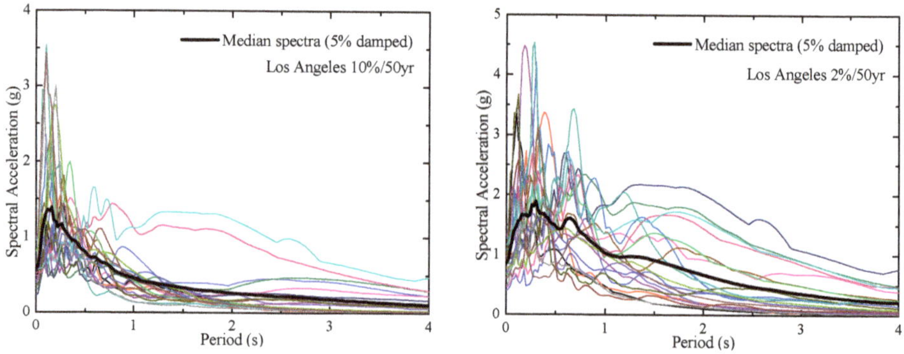

(**a**) la01-20 (**b**) la21-40

Figure 11. Response spectra in the Los Angeles region.

6. Seismic Risk Assessment of CFDST-BRW Structure

6.1. Prototype Structures

In this paper, the nine-story CFDST frame designed in Ref. [20] is used as the base model of the CFDST frame-BRW structure, as shown in Figure 12. For the CFDST frame-BRW structure, only a BRW is filled in the middle span, and the width of CFDST-BRW model is 3 m. The specific parameters are in Table 1.

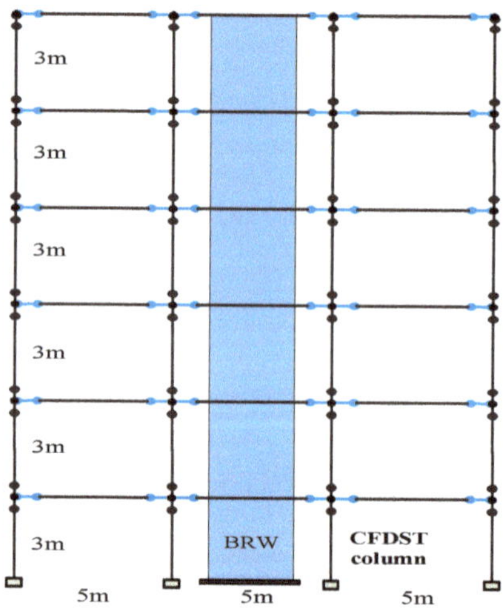

Figure 12. Schematic diagram of the CFDST frame-BRW structure.

Table 1. First-order natural vibration period of the BRW model.

Type	Size	T_1 (s)
CFDST -BRW	The thickness of reinforced concrete is 120 mm, and the spacing between steel bars is 206.130 mm. It is distributed in two layers along the thickness direction of the plate. The diameter of the steel bars is 10 mm. The thickness of the protective layer is 20 mm	1.415

The gravity load adopts the analysis example of Skalomenos [16], which is assumed to be $G + 0.3Q$ = 27.5 kN/m. The yield strength of the CFDST column and the steel strength in the BRW is 275 MPa. The compressive strength of concrete and filled concrete in the BRW is 20 Mpa.

6.2. Probabilistic Seismic Demands

The nonlinear dynamic analysis is carried out to establish the numerical analysis model. The selected representative seismic waves (la01-la40) are input into the model one by one according to 10%/50 year and 2%/50 year groups, and the seismic demand results of the model can be obtained. The $ISDA_{max}$ is used as a parameter to evaluate the structural performance.

The functional relationship between the $ISDA_{max}$ of each floor and the spectral acceleration S_a (T_1) [27] of the CFDST frame-BRW model obtained by nonlinear dynamic analysis is recorded as follows:

$$\theta_{max} = a s_a^b \varepsilon \tag{11}$$

where ε is a random variable; θ_{max} is the $ISDA_{max}$ obtained by NTHA; a and b are constants; S_a is the spectral acceleration corresponding to the first natural vibration period of the structure when the damping ratio is 5%.

6.3. Probabilistic Seismic Demands

The seismic vulnerability curve is the main part of structural seismic risk assessment, representing the functional relationship between the maximum dynamic response of the structure under an earthquake and the intensity of ground motion. The incremental dynamic analysis (IDA) of the CFDST frame-BRW is carried out using the selected seismic waves. Then, the curve relationship between the ground motion intensity and the dynamic response of each model is obtained. To reduce the calculation amount of dynamic analysis, each ground motion gradually increases the input ground motion intensity at a multiple of 100 gal and is gradually applied to the structure to record the relationship between the $ISDA_{max}$ and the input seismic intensity. The median value mC of the seismic demand of the three structural models under different performance states is obtained by counting these IDA curves. In addition, the seismic vulnerability curve also considers many uncertain factors, including the uncertainty of earthquakes, the uncertainty of structures, and the uncertainty of modeling, which correspond to the logarithmic standard deviations, including β_C for the uncertainty of earthquakes, β_{SC} for the uncertainty of structures and β_u for the uncertainty of modeling; β_{SC} can be combined with β_u to form $\beta_C = \sqrt{{\beta_{SC}}^2 + {\beta_u}^2}$. It is called the logarithmic standard deviation, corresponding to the uncertainty of seismic demand.

According to these median values and uncertainty parameters, the seismic vulnerability of each model to various performance states can be calculated by:

$$P[LS_i|S_a = x] = \Phi[\frac{(\ln ax^b/m_C)}{\sqrt{\beta_C^2 + \beta_D^2}}] \tag{12}$$

where the demand uncertainty β_D is determined by the dynamic response of a series of ground motions, which is equal to $\sigma_{ln\varepsilon}$; m_C is the median of earthquake demand obtained by IDA; and β_C is the logarithmic standard deviation of the earthquake demand group.

The seismic vulnerability of the CFDST frame-BRW model calculated in this paper only considers the influence of seismic uncertainty. It does not consider the impact of other uncertain factors, such as structural uncertainty and modeling uncertainty. These uncertainties need to be completed by carrying out many random numerical analyses. The authors will study the influence of the BRW on the seismic risk of CFDST frames. Therefore, only the uncertainty factors of earthquakes that have the greatest impact on them are selected. In the future, the authors will further study the seismic vulnerability and seismic risk of CFDST structures under multiple uncertainties and other damping structures [28].

Because multiple uncertainties are essential for accurately assessing structures' seismic risk, many studies have discussed this issue [29–38].

The CFDST-BRW model's seismic vulnerability is calculated using Equation (12). The fragility curves corresponding to each limit state in Figure 13 and the corresponding fragility parameters are shown in Table 2.

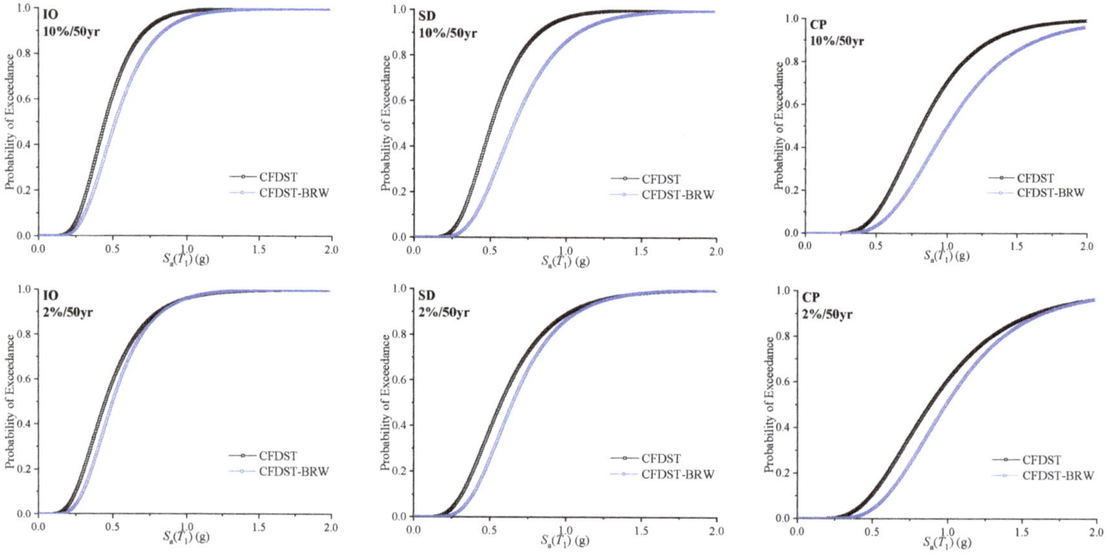

Figure 13. Fragility curves of the models associated with the LA ensemble.

Table 2. Fragility parameters for the models in the Los Angeles region.

Model	Earthquake Grouping	m_{IO}	m_{SD}	m_{CP}	β_R
CFDST-BRW	10%/50 year	0.224	0.390	1.021	0.382
	2%/50 year	0.208	0.377	0.983	0.391

6.4. Probability Estimation of Annual Performance Limits

The HAZUS software released by the USGS of the US Geological Survey was used to calculate the seismic disaster parameters of the model in the Los Angeles area of 10%/50 year and 2%/50 year earthquakes, as shown in Table 3. According to these earthquake disaster parameters, the shape parameter k and the proportional parameter k_0 of CFDST frame-BRW structures in Los Angeles can be obtained. The shape parameter k of the CFDST frame-BRW structure in this area is not largely different from the shape parameter k of the CFDST frame. The shape parameter k and the proportional parameter k_0 of the model are brought into the seismic risk function, and the seismic disaster curve of the model can be obtained, as shown in Figure 14. The annual average occurrence probability of the CFDST frame-BRW model can be calculated by convolution calculation of the seismic risk function and seismic vulnerability function. The results are shown in Table 4.

Table 3. Seismic hazard parameters for the Los Angeles region.

Model	T_1	S_a (T_1) (g)		k	k_0 $(\times 10^{-4})$
		10%/50 year	2%/50 year		
CFDST-BRW	1.415	0.522	0.785	4.051	1.489

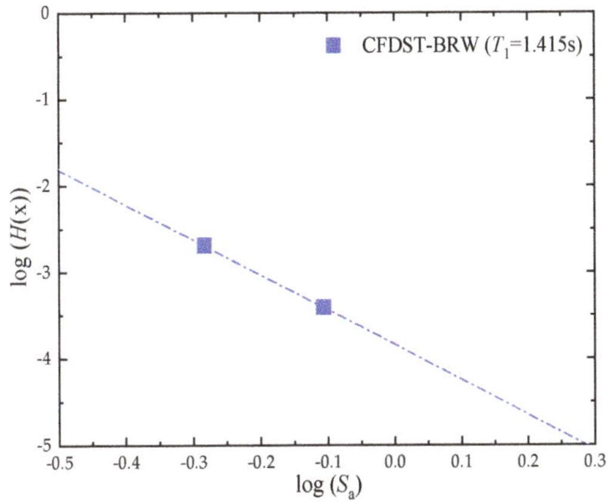

Figure 14. Seismic hazard curves for Los Angeles.

Table 4. Annual probability of limit state occurrence for the models in Los Angeles.

Structural Model	Seismic Category	$P\,[LS_i]$ ($\times 10^{-3}$)		
		IO	SD	CP
CFDST-BRW	10%/50 year	211.37	22.36	0.45
	2%/50 year	302.14	27.16	0.56

As shown in Figure 15, combined with Ref. [20], the BRW reduces the annual probability and collapse probability of the CFDST frame structure to achieve various performance states within the service life of 50 years. The BRW reduces the annual failure probability and 50-year collapse probability of the CFDST frame to a reasonable extent. The BRW reduces the 50-year collapse probability of the CFDST frame to 2.76%, significantly reducing the CFDST frame's collapse probability. The BRWs can effectively reduce the probability of different degrees of damage to the structure during the service life.

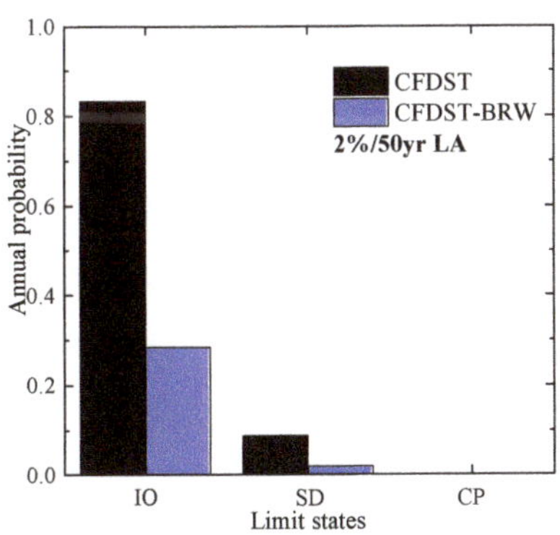

(**a**) Annual probability of occurrence

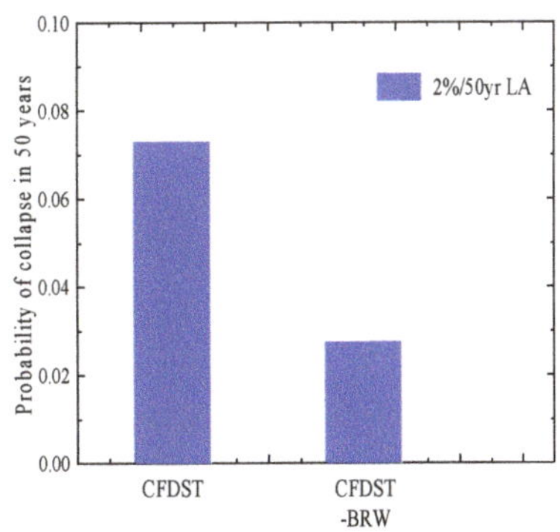

(**b**) The probability of collapse occurring in 50 years

Figure 15. Seismic risk probability of the CFDST frame shear wall model.

7. Conclusions

The seismic risk assessment of a nine-story CFDST frame-BRW structure was carried out to study the BRW on the seismic risk of the CFDST frame. The purpose was to calculate the exceedance probability of this kind of structure to reach various performance limit states during the design life period (usually 50 years). The ground motion obtained by the

SAC project represented the uncertainty of seismic demand. The following conclusions can be drawn.

1. Based on the unified theory, the model considered the mechanical properties of concrete materials under a three-dimensional stress state under compression. A simplified analysis model of a shear wall with beam-end connections of different types of materials was established. The hysteretic characteristics of a BRW under earthquake action were simulated by two nonlinear springs with Pinching04 material properties. The correctness of the adopted simplified analysis model was verified by the seismic performance test results of two CFDST frame-BRW structures.
2. A numerical analysis model of a nine-story CFDST frame-BRW structure was established, and ground motion records were selected from the SAC project to describe the seismic risks in the Los Angeles area of the US States, which were divided into two risk levels of 2%/50 years and 10%/50 years. This model uses nonlinear time history analysis and incremental dynamic analysis on the selected records.
3. The analysis model's fragility curves were established using three determined performance limits and structural responses. The annual probability of these three limits and the probability of collapse within 50 years were determined by convoluting the fragility with the seismic risk specified by the USGS.
4. The BRW reduced the probability of collapse of the CFDST frame to 2.76% in 50 years, which indicates the proposed BRW could effectively reduce the probability of different degrees of damage in the service cycle of the structure.

Author Contributions: Conceptualization, H.S., Y.H., J.Z. and Y.L.; Methodology, H.S.; Validation, H.S.; Investigation, H.S.; Resources, J.Z., D.W. and Y.L.; Data curation, H.S. and Y.H.; Writing—original draft, H.S. and Y.H.; Writing—review & editing, H.S. and Y.H.; Visualization, H.S. and Y.H.; Supervision, J.Z. and D.W.; Project administration, H.S. All authors have read and agreed to the published version of the manuscript.

Funding: This work was sponsored by the Hunan Natural Science Foundation of China (Grant No. 2021JJ41083) and the Hunan Provincial Department of Education Excellent Youth Fund (Grant No. 20B598).

Data Availability Statement: Not applicable.

Acknowledgments: The authors wish to thank the anonymous reviewers for their thorough evaluations and valuable comments, which have helped improve the paper.

Conflicts of Interest: The authors declare no conflict of interest.

References

1. Wang, X.D.; Gao, P.; Liu, J.P.; Zhao, Q.R.; Xu, H.M.; Wei, W.; Wang, X.T. Assessment on seismic performance of circular concrete-filled thin-walled steel tube columns. *Adv. Struct. Eng.* **2023**, *26*, 399–412. [CrossRef]
2. Skalomenos, K.A.; Hatzigeorgiou, G.D.; Beskos, D.E. Application of the hybrid force/displacement (HFD) seismic design method to composite steel/concrete plane frames. *J. Constr. Steel. Res.* **2015**, *115*, 179–190. [CrossRef]
3. Gan, D.; Zhou, Z.; Yan, F.; Zhou, X.H. Shear transfer capacity of composite sections in steel tubed-reinforced-concrete frames. *Structures* **2017**, *12*, 54–63. [CrossRef]
4. Lai, Z.C.; Huang, Z.H.; Varma, A.H. Seismic analysis and performance of high strength composite special moment frames (C-SMFs). *Structures* **2017**, *9*, 165–178. [CrossRef]
5. Wang, K.; Yuan, S.F.; Chen, Z.X.; Zhi, H.X.; Shi, G.L.; Cao, D.F. Experimental study on hysteretic behavior of composite frames with concrete-encased CFST columns. *J. Constr. Steel. Res.* **2016**, *123*, 110–120. [CrossRef]
6. Bai, Y.T.; Hu, H.S.; Wang, J.T.; Sun, Q. Modeling on collapse behaviour of high-rise concrete-filled steel composite frames under over-design seismic excitations. *Struct. Infrastruct. Eng.* **2017**, *13*, 1563–1575. [CrossRef]
7. Ren, F.M.; Zhou, Y.; Zhang, J.B.; Lin, S.M. Experimental study on seismic performance of CFST frame structures with energy dissipation devices. *J. Constr. Steel. Res.* **2013**, *90*, 120–132. [CrossRef]
8. Jia, M.M.; Lu, D.G.; Guo, L.H.; Sun, L. Experimental research and cyclic behavior of buckling-restrained braced composite frame. *J. Constr. Steel. Res.* **2014**, *95*, 90–105. [CrossRef]
9. Nie, J.G.; Hu, H.S.; Fan, J.S.; Tao, M.X.; Li, S.Y.; Liu, F.J. Experimental study on seismic behavior of high-strength concrete filled double-steel-plate composite walls. *J. Constr. Steel. Res.* **2013**, *88*, 206–219. [CrossRef]

10. Wang, J.F.; Li, B.B. Cyclic testing of square CFST frames with ALC panel or block walls. *J. Constr. Steel. Res.* **2017**, *130*, 264–279. [CrossRef]
11. Guo, L.H.; Rong, Q.; Qu, B.; Liu, J.P. Testing of steel plate shear walls with composite columns and infill plates connected to beams only. *Eng. Struct.* **2017**, *136*, 165–179. [CrossRef]
12. Hu, Y.; Zhao, J.H.; Zhang, D.F.; Li, Y.P. Cyclic performance of concrete-filled double-skin steel tube frames strengthened with beam-only-connected composite steel plate shear walls. *J. Build. Eng.* **2020**, *31*, 15. [CrossRef]
13. Bian, J.L.; Cao, W.L.; Dong, H.Y.; Chen, J.W. Seismic research of prefabricated CFST frame with various H-steel frame shear walls. *J. Build. Eng.* **2022**, *57*, 20. [CrossRef]
14. Hu, Y.; Zhao, J.H.; Zhang, D.F.; Zhang, H.Q. Cyclic tests of fully prefabricated concrete-filled double-skin steel tube/moment-resisting frames with beam-only-connected steel plate shear walls. *Thin-Walled Struct.* **2020**, *146*, 14. [CrossRef]
15. Xu, S.Y.; Li, Z.L.; Liu, H.J. Global seismic performance of a new precast CFST column to RC beam braced frame: Shake table test and numerical study. *Steel Compos. Struct.* **2016**, *21*, 805–827. [CrossRef]
16. Skalomenos, K.A.; Hatzigeorgiou, G.D.; Beskos, D.E. Modeling level selection for seismic analysis of concrete-filled steel tube/moment-resisting frames by using fragility curves. *Earthq. Eng. Struct. Dyn.* **2015**, *44*, 199–220. [CrossRef]
17. Leon, R.T.; Gao, Y. Resiliency of steel and composite structures. *Front. Struct. Civ. Eng.* **2016**, *10*, 239–253. [CrossRef]
18. Kamaris, G.S.; Skalomenos, K.A.; Hatzigeorgiou, G.D.; Beskos, D.E. Seismic damage estimation of in-plane regular steel/concrete composite moment resisting frames. *Eng. Struct.* **2016**, *115*, 67–77. [CrossRef]
19. Ahmadi, M.; Naderpour, H.; Kheyroddin, A.; Gandomi, A.H. Seismic failure probability and vulnerability assessment of steel-concrete composite structures. *Period. Polytech.-Civ. Eng.* **2017**, *61*, 939–950. [CrossRef]
20. Hu, Y.; Zhao, J.H.; Zhang, D.F.; Zhang, Y.F. Seismic risk assessment of concrete-filled double-skin steel tube/moment-resisting frames. *Earthq. Struct.* **2018**, *14*, 249–259. [CrossRef]
21. FEMA. *FEMA 356 Prestandard and Commentary for the Seismic Rehabilitation of Buildings*; FEMA: Washington, DC, USA, 2000.
22. HAZUS. *HAZUS Earthquake Loss Estimation Methodology. National Institute of Building for the FEMA*; HAZUS: Washington, DC, USA, 1997.
23. SEAOC. *SEAOC Appendix I: Tentative Guidelines for Performance-Based Seismic Engineering*; SEAOC: Sacramento, CA, USA, 1999.
24. Ibarra, L.F.; Medina, R.A.; Krawinkler, H. Hysteretic models that incorporate strength and stiffness deterioration. *Earthq. Eng. Struct. Dyn.* **2005**, *34*, 1489–1511. [CrossRef]
25. Muhummud, T. Seismic Behavior and Design of Composite SMRFs with Concrete Filled Steel Tubular Columns and Steel Wide Flange Beams. Ph.D. Dissertation, Lehigh University, Bethlehem, PA, USA, 2023.
26. Lignos, D.G.; Krawinkler, H. Deterioration modeling of steel components in support of collapse prediction of steel moment frames under earthquake loading. *J. Struct. Eng.* **2011**, *137*, 1291–1302. [CrossRef]
27. Ellingwood, B.R.; Kinali, K. Quantifying and communicating uncertainty in seismic risk assessment. *Struct. Saf.* **2009**, *31*, 179–187. [CrossRef]
28. Soni, D.P.; Mistry, B.B.; Jangid, R.S.; Panchal, V.R. Seismic response of the double variable frequency pendulum isolator. *Struct. Control Health Monit.* **2011**, *18*, 450–470. [CrossRef]
29. Dolsek, M. Simplified method for seismic risk assessment of buildings with consideration of aleatory and epistemic uncertainty. *Struct. Infrastruct. Eng.* **2012**, *8*, 939–953. [CrossRef]
30. Tubaldi, E.; Barbato, M.; Ghazizadeh, S. A probabilistic performance-based risk assessment approach for seismic pounding with efficient application to linear systems. *Struct. Saf.* **2012**, *36–37*, 14–22. [CrossRef]
31. Tubaldi, E.; Barbato, M.; Dall'Asta, A. Performance-based seismic risk assessment for buildings equipped with linear and nonlinear viscous dampers. *Eng. Struct.* **2014**, *78*, 90–99. [CrossRef]
32. Kosic, M.; Fajfar, P.; Dolsek, M. Approximate seismic risk assessment of building structures with explicit consideration of uncertainties. *Earthq. Eng. Struct. Dyn.* **2014**, *43*, 1483–1502. [CrossRef]
33. Asgarian, B.; Ordoubadi, B. Effects of structural uncertainties on seismic performance of steel moment resisting frames. *J. Constr. Steel. Res.* **2016**, *120*, 132–142. [CrossRef]
34. Melani, A.; Khare, R.K.; Dhakal, R.P.; Mander, J.B. Seismic risk assessment of low rise RC frame structure. *Structures* **2016**, *5*, 13–22. [CrossRef]
35. Kechidi, S.; Macedo, L.; Castro, J.M.; Bourahla, N. Seismic risk assessment of cold-formed steel shear wall systems. *J. Constr. Steel. Res.* **2017**, *138*, 565–579. [CrossRef]
36. Jalayer, F.; Ebrahimian, H. Seismic risk assessment considering cumulative damage due to aftershocks. *Earthq. Eng. Struct. Dyn.* **2017**, *46*, 369–389. [CrossRef]
37. Aghababaei, M.; Mahsuli, M. Component damage models for detailed seismic risk analysis using structural reliability methods. *Struct. Saf.* **2019**, *76*, 108–122. [CrossRef]
38. Lu, D.G.; Yu, X.H.; Jia, M.M.; Wang, G.Y. Seismic risk assessment for a reinforced concrete frame designed according to Chinese codes. *Struct. Infrastruct. Eng.* **2014**, *10*, 1295–1310. [CrossRef]

buildings

MDPI

Article

Seismic Fragility Analysis of Existing RC Frame Structures Strengthened with the External Self-Centering Substructure

Weiheng Liu [1], Jianwei Zhang [1], Hang Liu [2,*], Fei Wang [1], Juan Liu [3] and Mingjie Han [2]

1 Key Laboratory of Urban and Engineering Security Disaster of China Ministry of Education, Beijing University of Technology, Beijing 100124, China
2 Beijing Building Construction Research Institute Co., Ltd., Beijing 100039, China
3 Department of Civil Engineering, Tsinghua University, Beijing 100084, China
* Correspondence: liuhang71@sina.com

Abstract: Based on a practical engineering case of seismic strengthening, this paper used the enlarging cross-section method and an external self-centering substructure to improve the seismic performance and seismic resilience of existing frame structures. Among them, the external self-centering substructure included setting a self-centering precast beam and diagonal braces. Utilizing the OpenSees finite element platform, a seismic fragility analysis was carried out to compare the improvements in seismic performance and seismic resilience before and after strengthening. The analysis results show that the proposed modelling method could be simulated satisfactorily. The maximum inter-story drift and the residual inter-story drift of the strengthened frame structures decreased significantly under the same peak ground acceleration. The peak ground acceleration of the strengthened frame structures significantly increased under different performance levels. Additionally, the exceedance probability of the strengthened frame structures was obviously reduced, which reflected that the seismic performance and seismic resilience of the strengthened frame structures were significantly improved.

Keywords: frame structures; seismic strengthening; numerical simulation; IDA; seismic fragility analysis

Citation: Liu, W.; Zhang, J.; Liu, H.; Wang, F.; Liu, J.; Han, M. Seismic Fragility Analysis of Existing RC Frame Structures Strengthened with the External Self-Centering Substructure. *Buildings* **2023**, *13*, 2117. https://doi.org/10.3390/buildings13082117

Academic Editors: Liqiang Jiang, Wei Chen, Chang He, Yi Hu, Qi Cai and Francisco López-Almansa

Received: 23 July 2023
Revised: 9 August 2023
Accepted: 16 August 2023
Published: 21 August 2023

1. Introduction

As economic development enters a new phase, the growth rate of infrastructure construction is gradually slowing down. Besides, the number of new buildings is also decreasing year by year due to limited urban land and space. A large number of early RC frame structures exist in cities. Although they have not reached the end of their design service life, there are still some potential problems. Firstly, it is difficult for early RC multi-story frame structures to meet the requirements of the current Code for the Seismic Design of Buildings (GB 50011-2010) [1] due to the rapid development of seismic technology. Secondly, the function of the existing frame structures may be changed during their service life. For example, some frame columns may be removed in order to open up large amounts of space; this means the lateral stiffness of frame structures is greatly reduced, which then presents a difficult problem for strengthening existing frame structures. Thirdly, early and existing frame structures are demolished, and new frame structures are built, which incurs high costs and other problems. In summary, the seismic strengthening of existing frame structures has gradually become a prominent issue in research.

There are various strengthening methods [2–4], and a simple and common method is the enlarging cross-section method. Existing frame structures can be strengthened by increasing the size of the concrete cross-section and the amount of steel bars to improve the lateral stiffness and bearing capacity. Additionally, the enlarging cross-section method could effectively improve the seismic performance and collapse resistance of existing frame structures under strong earthquakes [5,6]. However, the enlarging cross-section method has limited effects on reducing residual deformation. According to a performance-based seismic design method, prestress was applied to steel strands so that its residual deformation was reduced and its seismic resilience was increased, which was easy to repair after earthquakes [7]. In addition, the two ends of the steel strand were anchored to the cast-in-place concrete, and the performance of the co-working was good. Priestley et al. [8,9] studied the precast seismic structural systems (PRESSS) program and proposed a precast concrete frame structure system with dry prestressed hybrid connection joints. Lu et al. [10,11] conducted seismic performance tests on self-centering RC frame structures with beam–column joints and column–base joints. It was found that the frame structures were almost undamaged under extreme earthquake action, and the seismic performance and self-centering capacity were excellent. Kurosawa et al. [12] used precast prestressed concrete frames to retrofit existing RC frame structures, and found that precast frames had well-controlled cracking and minimized residual deformation compared to integral RC frames. Eldin et al. [13] used post-tensioned prestressed technology to retrofit existing frame structures, and the results showed that the maximum inter-story drift and residual inter-story drift of strengthened frame structures were significantly reduced. It could be seen that the use of prestressed steel strands effectively enhanced the seismic performance and seismic resilience of the frame structures.

In order to scientifically and reasonably evaluate the seismic performance and seismic resilience of frame structures before and after strengthening, the seismic fragility analysis has been developed [14]. The seismic fragility analysis, which quantitatively describes the dynamic relationship between the peak ground acceleration and exceedance probability of frame structures, plays an important role in global seismic risk assessment [15]. In addition, the seismic fragility analysis is relatively easy to perform, and the results are relatively intuitive. More and more scholars have adopted the seismic fragility analysis to assess the seismic performance and seismic resilience of structures in recent years. Ji et al. [16] proposed a simple lumped-parameter model for the seismic fragility assessment of existing high-rise buildings. Sarno et al. [17,18] conducted a seismic fragility analysis considering the influence of reinforcement corrosion on existing RC frame structures. Kumar et al. [19,20] considered the uncertainty in aleatory and epistemic sources, and evaluated the seismic fragility of existing low-, medium-, and high-rise RC frame structures. Pitilakis et al. [21] proposed a modular approach for a large-scale seismic fragility assessment of existing RC frame structures, including soil–structure interactions and site amplification effects. Gautam et al. [22] conducted a seismic fragility analysis of RC frame structures with structural and non-structural components, and found that the global fragility function could be used for overall loss estimation, but the component level fragility functions could be used to identify weak links in frame structures. Palagala et al. [23] proposed a quick seismic fragility assessment technique for existing RC frame structures, providing qualitative and quantitative assessments in a more engineered way. Dalal et al. [24] proposed a performance-based plastic design method for a seismic fragility analysis of RC frame structures, and found that the designed frame structure met three performance levels, namely immediate occupancy, life safety and collapse prevention. Cao et al. [25] proposed a new strengthening method for existing frames by using external substructures, and conducted incremental dynamic analysis (IDA) and seismic fragility analysis. The results illustrated that the strengthened RC frames had a higher bearing capacity and lower structural damage. It could be seen that the seismic fragility analysis was a relatively effective method for evaluating RC frame structures, and also contributed to achieving performance-based seismic design goals.

In brief, the finite element analysis was used to simulate the strengthened existing frame structures. On this basis, the dynamic increment analysis and seismic fragility analysis were conducted to evaluate its seismic performance and seismic resilience on a pilot project. The analysis results could be used for performance-based design, and could provide certain design recommendations.

2. Basic Information of RC Frame Structures

The basic information of the RC frame structures was introduced before and after seismic strengthening, including geometric dimensions, reinforcement layout, and load layout, etc.

2.1. Frame KJ1

The pilot project of seismic strengthening was a four-storey RC frame structure (namely frame KJ1), as shown in Figure 1. The total area of the frame KJ1 was approximately 1900 m^2. The length was 25.5 m in the east–west direction, and the width was 18.0 m in the north–south direction. The total height of the frame KJ1 was 14.8 m, and the storey heights from the ground floor to the top floor were 4.11 m, 3.30 m, 3.30 m and 4.09 m, respectively. The RC frame structure was located in a seismic precautionary intensity of 8 degrees. The site class was Class II. The design earthquake group was Group II. Additionally, the characteristic period value was 0.4 s.

Figure 1. Scene photos of frame KJ1.

The structural plan layout of frame KJ1 and the cross-sectional dimensions of beams and columns are depicted in Figure 2.

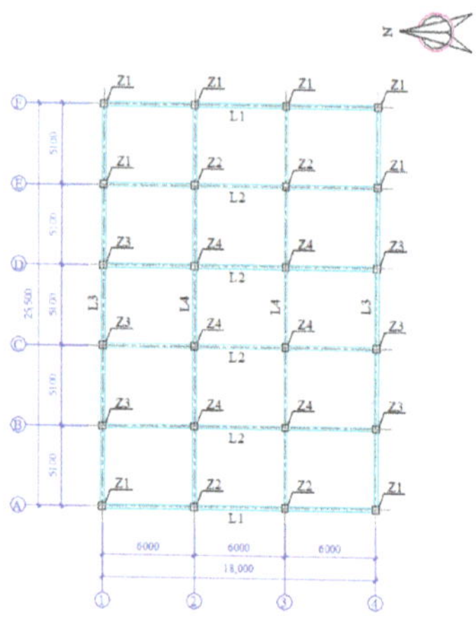

(a)

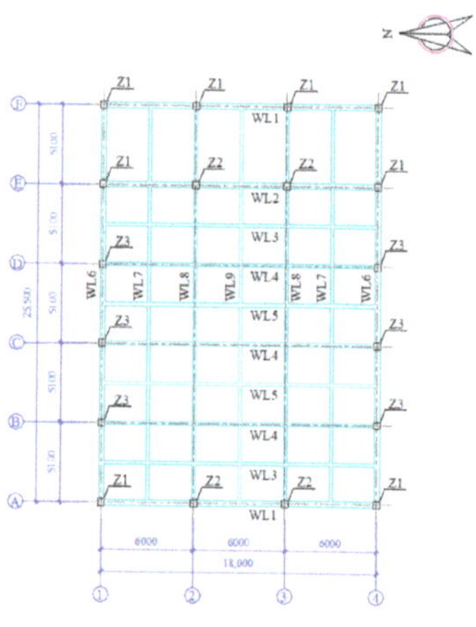

(b)

Figure 2. *Cont.*

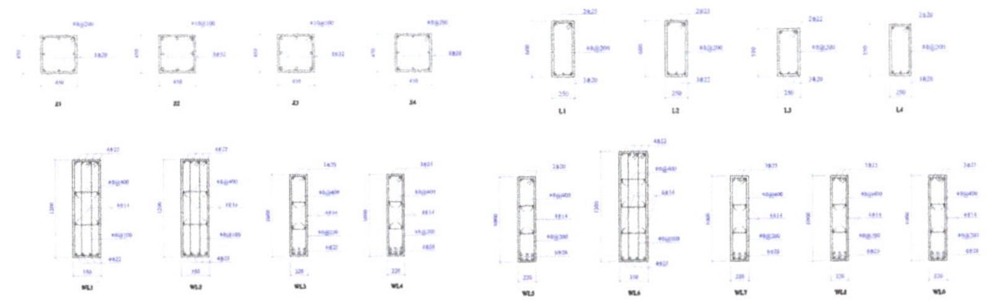

(c)

Figure 2. Structural plan layout of frame KJ1. (**a**) Plan layout of the first to the third storey; (**b**) Plan layout of the fourth storey; (**c**) Cross-sectional dimensions of beams and columns.

The plan layout of the line load on the beam, the floor dead load and the floor live load is presented in Figure 3. The floor uniform load was marked at the center of the floor. Among them, the number outside the brackets indicated the floor dead load, and the number inside the brackets indicated the floor live load.

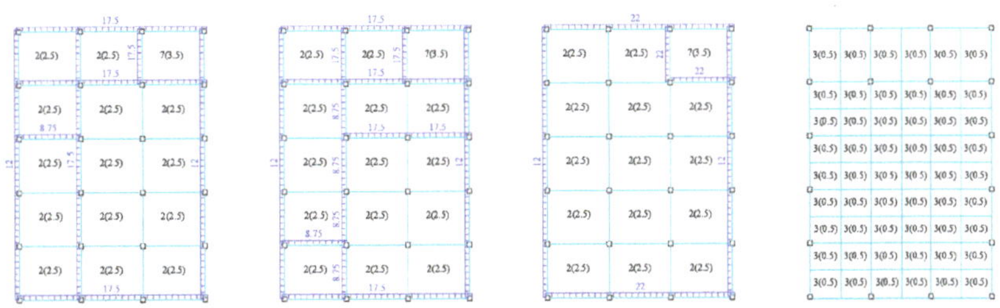

(**a**) (**b**) (**c**) (**d**)

Figure 3. Load plan layout of frame KJ1. (**a**) First storey; (**b**) Second storey; (**c**) Third storey; (**d**) Fourth storey.

2.2. Frame KJ2

According to the requirements of first party, it was planned to remove the four frame columns from the first to the third storey. The interior of the frame KJ1 was changed to a large space, which resulted in a significant reduction in the lateral stiffness. The bearing capacity and deformation capacity of the frame structure did not meet the code requirements. According to the performance-based seismic design method, the original frame structure was strengthened by the enlarging cross-section method and the external self-centering substructure to improve the seismic resilience, namely frame KJ2. The self-centering precast beams were set up on the north–south exterior facades, and the diagonal braces were set up on the west–east exterior facades. Additionally, the diameter of the diagonal braces was taken as 50 mm. The structural plan layout of frame KJ2 and the cross-sectional dimensions of the beams and columns are plotted in Figure 4.

After strengthening the frame KJ1 by the enlarging cross-section method, the floor dead load increased, while the line load on beam and the floor live load remained unchanged. The load plan layout of frame KJ2 is depicted in Figure 5.

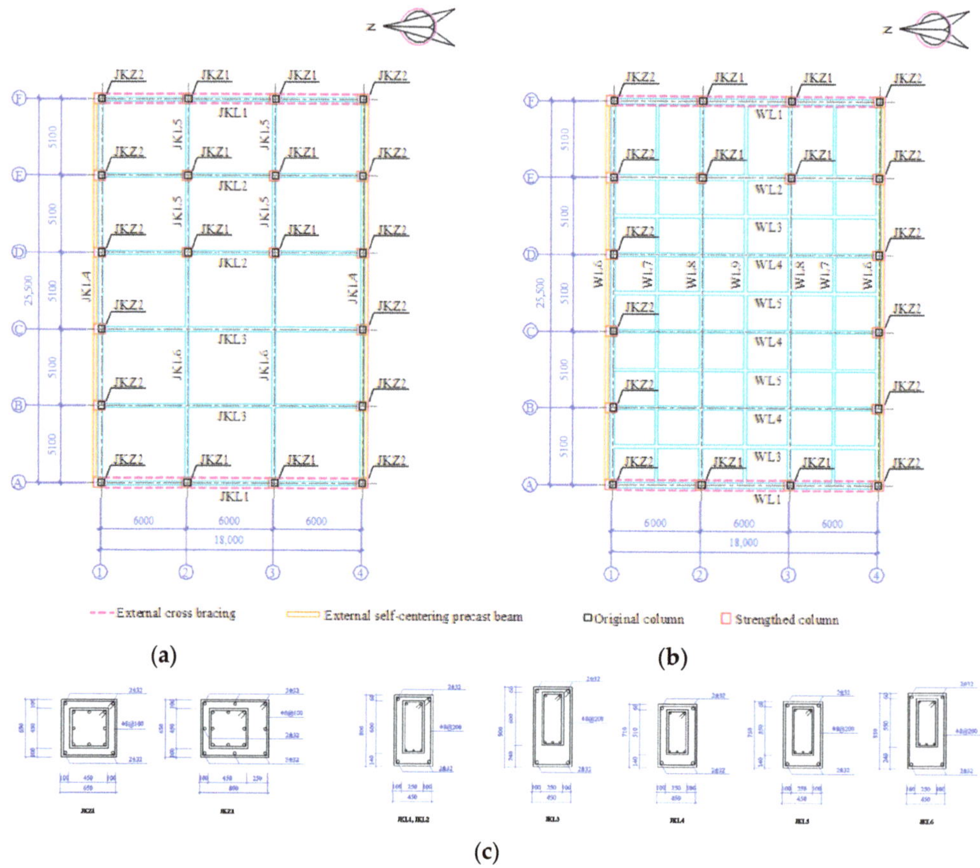

Figure 4. Structural plan layout of frame KJ2. (**a**) Plan layout of the first to the third storey; (**b**) Plan layout of the fourth storey; (**c**) Cross-sectional dimensions of beams and columns.

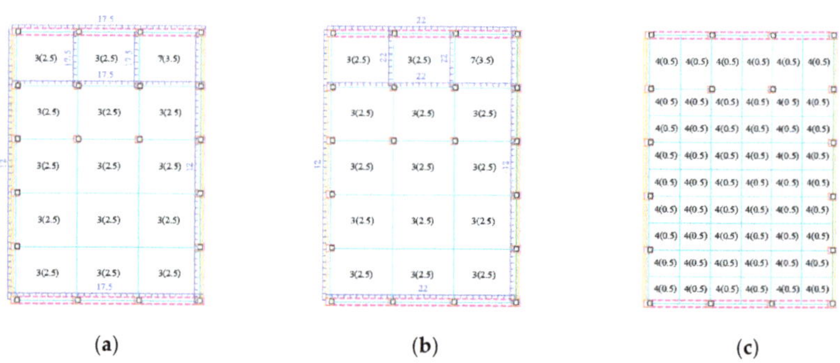

Figure 5. Load plan layout of frame KJ2. (**a**) First to second storey; (**b**) Third storey; (**c**) Fourth storey.

3. Establishment and Verification of Finite Element Models

OpenSees (Open System for Earthquake Engineering Simulation) (University of California, Berkeley, CA, USA) is an open-source seismic engineering numerical simulation platform with a high computational efficiency and sustainable development rate. Firstly, the simulated hysteresis curve of one-story one-span frame structures were compared with quasi-static test results using the OpenSees version 2.5.0 as the numerical simulation platform. Secondly, the whole finite element models of the frame structures were established, and the incremental dynamic analysis and the seismic fragility analysis were carried out.

3.1. Finite Element Modelling Methods

The fiber model of OpenSees was chosen to establish the finite element model. Additionally, the frame beams and columns were simulated by beam–column elements. Concrete02 material was used to simulate the concrete in the frame structure and precast beams, taking into account the tensile strength and changes in loading and unloading stiffness of the concrete. The constitutive relationship of the core concrete adopted the Mander model [26] considering the constraint effect of stirrups. The Steel02 material was used for simulating the normal steel bars, energy-dissipating bars and steel strands in the frame structures. The steel strand adopted the Truss element to achieve an unbonded effect with concrete. In addition, it should be noted that the steel02 material can be prestressed, and the prestress of 250 MPa and 450 MPa was applied to steel strands in the self-centering precast beam and diagonal braces, respectively.

The connection part between the precast beam and the frame structure was a challenging part of the numerical simulation. The precast beams could be rotated in their plane. Additionally, the connection of the precast beam and the frame structure relied on steel strands and energy-dissipating bars. The ZeroLengthSection was set at the connection and given the ENT (Elastic-No Tension) material. Moreover, energy-dissipating bars were placed in the ZeroLengthSection. Considering the deformation compatibility between the precast beam and the frame structure during the loading process, the multi-point constraint (equalDOF) command was used to constrain the degrees of freedom for the nodes. Three nodes were constrained to the same position. Node 3 was the frame joint, and Node 2 was the end of the steel strand. The degrees of freedom for Node 2 and Node 3 were exactly the same. Node 1 was the end of the self-centering precast beam, and could rotate in the plane. The specific modelling method of self-centering precast beams is shown in Figure 6.

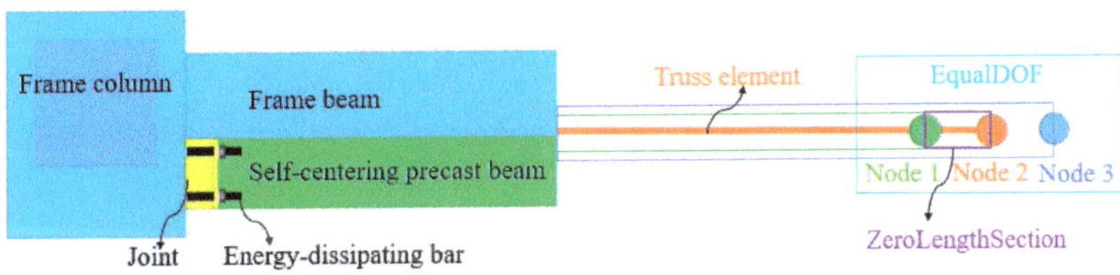

Figure 6. Modelling methods of self-centering precast beams.

3.2. One-Story One-Span Frame Structures

In order to verify the finite element modeling method, the quasi-static test results of one-story, one-span frame structures were compared with the finite element simulation results.

3.2.1. Test Overview

The schematic diagrams and cross-section reinforcement of one-story, one-span frame structures are plotted in Figure 7, and detailed information refers to references [27,28]. K1 and K2 represented two frame structures of different beam spans strengthened with diagonal braces and self-centering precast beams, respectively. The final number 1 indicated the original frame structure, and number 2 indicated the strengthened frame structure. The column height was 2400 mm, and the cross-sectional dimensions of columns for K1-1 and K2-1 were 270 × 270 mm. The beam spans of K1-1 and K2-1 were 3600 mm and 3060 mm, respectively. The beam heights of K1-1 and K2-1 were 360 mm and 306 mm, respectively. Additionally, the beam widths of K1-1 and K2-1 were 150 mm. Compared with K1-1, K1-2 was strengthened by the enlarging cross-section method and setting the diagonal braces, and the cross-sectional dimensions of columns for K1-2 were 390 × 390 mm. Compared with K2-1, K2-2 was strengthened by the enlarging cross-section method and setting the self-centering precast beam. The cross-sectional dimensions of columns for K2-2 were 390 × 480 mm. The beam heights of K1-2 and K2-2 were 480 mm, and the beam widths of K1-2 and K2-2 were 270 mm.

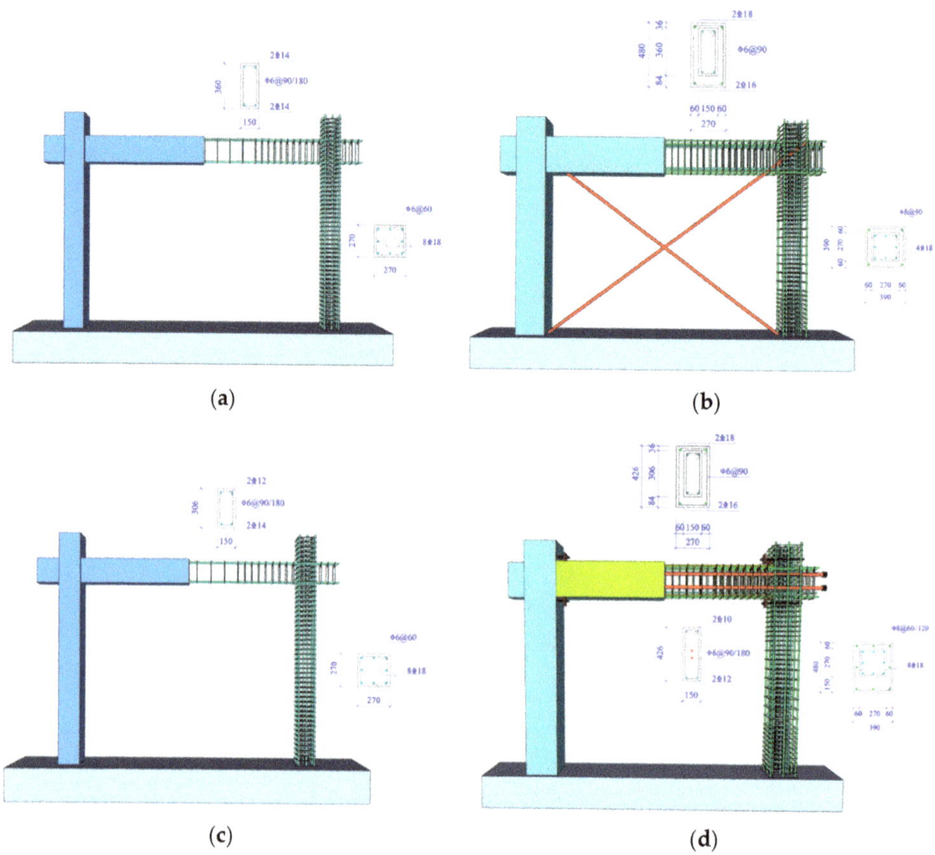

Figure 7. Schematic diagram and cross-section reinforcement of one-story one-span frame structures. (**a**) K1-1; (**b**) K1-2; (**c**) K2-1; (**d**) K2-2.

For K1-2, the bottom of steel strands was the fixed end, and the top of steel strands was the tensioned end. Among them, the tensioned end used wedge-type anchorage, and the fixed end used extruding anchorage, as displayed in Figure 8a. For K2-2, both ends of steel strands were simultaneously tensioned before the test, as depicted in Figure 8b. The energy-dissipating bars were tapped and then fixed with nuts.

The origin concrete strength grade of K1-1 and K2-1 was C20, and the post-cast concrete strength grade of enlarging cross-section was C40. The mechanical performance index of concrete is presented in Table 1, and the mechanical performance index of steel is listed in Table 2.

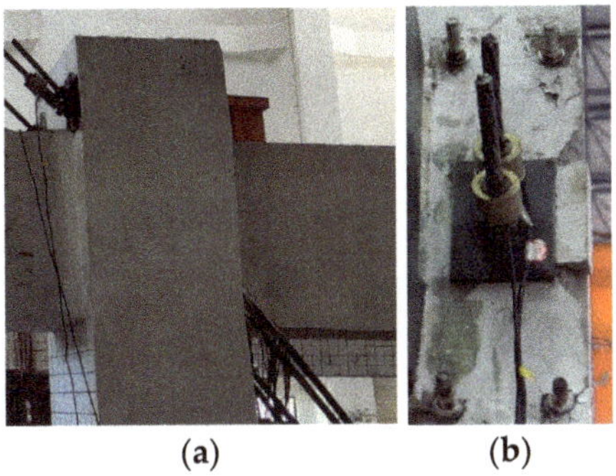

Figure 8. Scene photos of anchorage for steel strands. (**a**) K1-2; (**b**) K2-2.

Table 1. Mechanical performance index of concrete.

Concrete Strength Grade	f_{cu}/MPa	E_c/($\times 10^4$ N/mm^2)
C20	25.0	2.80
C40	39.3	3.25

Table 2. Mechanical performance index of steel.

Steel Type	f_y/MPa	E_s/($\times 10^5$ N/mm^2)
A6	293	2.05
A8	298	2.05
C12	497	2.01
C14	443	2.01
C18	459	2.00
A^S12.5	1680	1.95

3.2.2. Model Establishment of One-Story One-Span Frame Structures

Based on the quasi-static test, the above modelling method was adopted. According to the actual cross-sectional size and reinforcement information, finite element models of one-story one-span frame structures were established, as illustrated in Figure 9.

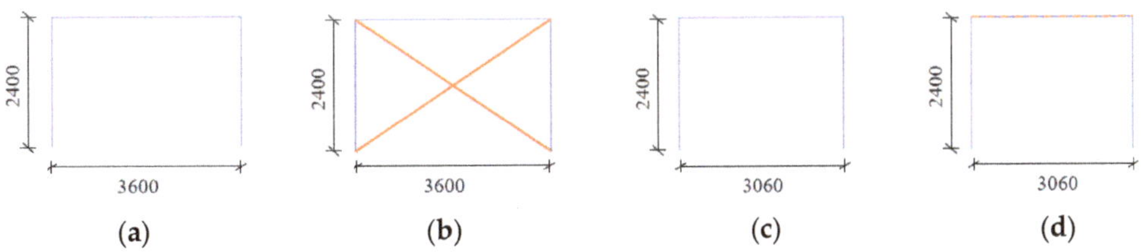

Figure 9. Finite element models of one-story one-span frame structures. (**a**) K1-1 (**b**) K1-2 (**c**) K2-1 (**d**) K2-2.

3.2.3. Model Verification of One-Story One-Span Frame Structures

The hysteresis curves of the test and simulation are plotted in Figure 10. It could be seen that the test results were generally in good agreement with the finite element simulation results. All errors were within 20%, confirming the rationality and reliability of the modelling method.

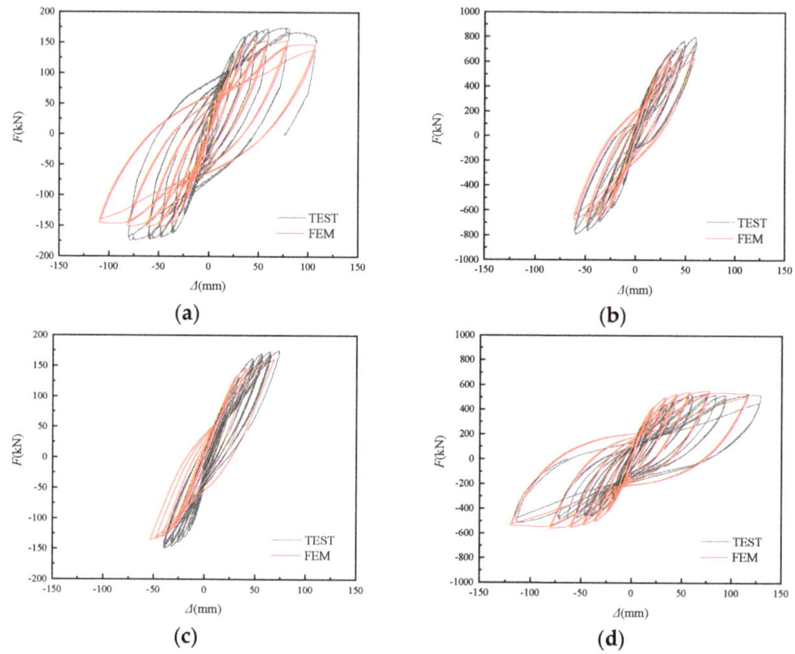

Figure 10. Comparisons of test and simulation results for one-story one-span frame structures. (**a**) K1-1; (**b**) K1-2; (**c**) K2-1; (**d**) K2-2.

3.3. RC Frame Structures

According to finite element simulation of one-story, one-span frame structures, the finite element models of four-storey frame structures were established, and the correctness of the finite element models was verified.

3.3.1. Model Establishment of RC Frame Structures

According to the actual size and load plan layout of RC frame structures, finite element models were established, as displayed in Figure 11.

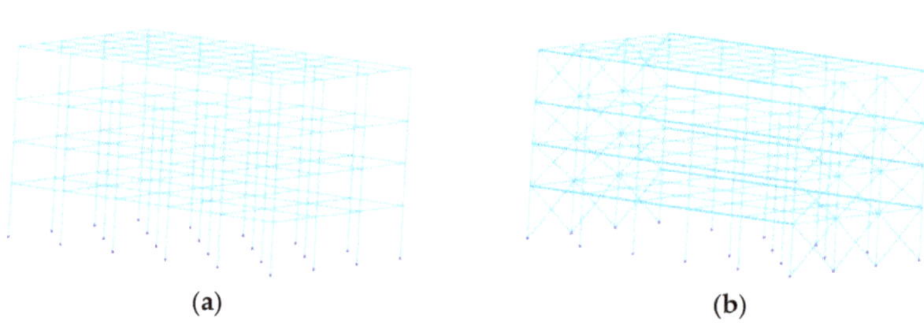

Figure 11. Finite element models of RC frame structures. (**a**) KJ1; (**b**) KJ2.

3.3.2. Verification of RC Frame Structures

The finite element models of the RC frame structures were established by YJK and OpenSees, respectively. The modal analysis of the RC frame structures was conducted to compare the natural vibration periods of the first three modes to verify the rationality of the finite element modelling method. The comparison of the natural vibration periods is exhibited in Table 3. The results showed that the natural vibration periods of the first three modes were basically similar. The errors were less than 20%, which indicated that the model's establishment of the frame structures was reasonable within this paper.

Table 3. Comparison of natural vibration periods.

Frame Number	Calculation Software	T1/s	T2/s	T3/s
KJ1	YJK	1.159	1.143	1.015
	OpenSees	0.966	0.954	0.869
KJ2	YJK	0.511	0.443	0.382
	OpenSees	0.498	0.487	0.389

4. Earthquake Ground Motion Input

Twenty earthquake ground motion records were selected according to the natural vibration period, damping ratio, site characteristic period, site class and design earthquake groups, etc. Detailed information of earthquake ground motion records is summarized in Table 4. The response spectrum analysis of earthquake ground motion records was conducted as plotted in Figure 12. Additionally, the average response spectrum was compared with the standard response spectrum of a seismic precautionary intensity of 8 degrees. The error between each earthquake response spectrum and the standard response spectrum was within 20% near the first period of frame structures.

Table 4. Earthquake ground motion records.

No.	Database Code	Year	Magnitude	PGA/Gal	Duration Time/s	Time Interval/s
1	NGA_no_40_A-SON033	1968	6.63	40.252	39.995	0.005
2	NGA_no_366_H-VC6090	1983	6.36	74.745	39.990	0.010
3	NGA_no_1010_5082A-325	1994	6.69	247.560	55.325	0.005
4	NGA_no_1184_CHY010-W	1999	7.62	222.011	131.996	0.004
5	NGA_no_2676_TTN024-V	1999	6.20	4.225	34.995	0.005
6	NGA_no_2721_CHY057-N	1999	6.20	24.109	49.995	0.005
7	NGA_no_2787_HWA039-V	1999	6.20	15.959	48.995	0.005
8	NGA_no_2914_TTN018-N	1999	6.20	10.057	42.995	0.005
9	NGA_no_3160_TCU014-N	1999	6.20	14.362	61.990	0.005
10	NGA_no_3291_CHY061-N	1999	6.30	27.555	50.990	0.005
11	NGA_no_3454_TCU046-N	1999	6.30	25.643	47.945	0.005
12	NGA_no_3485_TCU095-V	1999	6.30	16.870	66.995	0.005
13	000538ZA	1992	5.48	18.031	33.140	0.010
14	001967YA	1985	5.50	8.349	13.080	0.010
15	006968ZA	1999	6.20	3.267	33.780	0.010
16	AKT016908222_M	2017	-	95.404	95.000	0.020
17	AKT0130807240026NS	2008	6.80	23.963	174.000	0.010
18	AKTH01618222_M	2019	-	44.269	204.000	0.020
19	AKTH02311144_M	2011	-	70.392	300.000	0.020
20	SZO0190908110507EW	2009	6.50	96.006	138.000	0.010

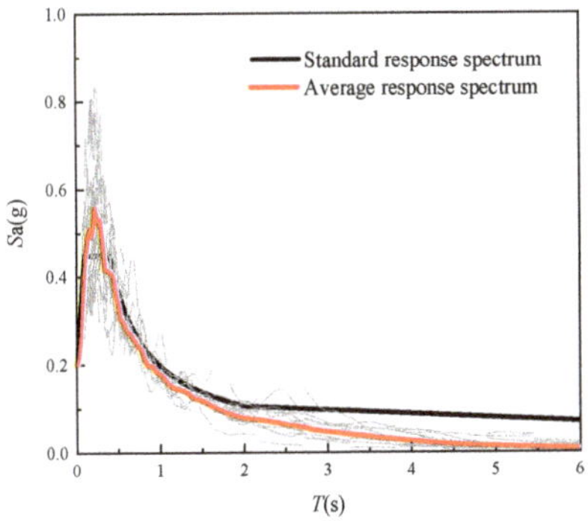

Figure 12. Response spectrum analysis of earthquake ground motion records.

5. Incremental Dynamic Analysis of RC Frame Structures

Incremental dynamic analysis (IDA) [29] is a parameter analysis method that comprehensively evaluates the performance level of frame structures under earthquake action. Besides, the IDA curve better reflects the evolution process of the structures from intact to failure. The specific operation process is that several earthquake ground motion records are applied to the structure model, and then each earthquake ground motion record is scaled to several intensity levels, finally obtaining a series of response curves. The peak ground acceleration (a_{PG}) was selected as the intensity measure (IM), and the maximum inter-story drift (θ_{max}) was selected as the damage measure (DM). The peak ground acceleration underwent a total of 34 amplitude modulations during the IDA. The amplitude modulation increments of 0.1~1.0 g were 0.1 g; the amplitude modulation increments of 1.2~5.0 g were 0.2 g; and the amplitude modulation increments of 5.5~7.0 g were 0.5 g. In order to reduce the computational complexity and reflect various performance points on the IDA curves based on different seismic demand parameters, the ultimate state of IDA analysis was the point where the structure reached dynamic instability, the nonlinear time history analysis did not converge, the maximum inter-story drift reached 10%, or the slope reached the initial slope of 20% in the IDA curve. Additionally, the point with the smallest value was selected as the ultimate state of collapse.

Taking the No. 1 earthquake ground motion record as an example, the maximum inter-story drift, the residual inter-story drift, and IDA were carried out. Then, twenty earthquake ground motion records were input, and IDA curves and fractile curves were obtained through statistical analysis. The calculation results and analyses were as follows.

5.1. Distribution Diagram of Maximum Inter-Story Drifts

FEMA 356 [30] defines three performance points of the maximum inter-story drift (θ_{max}). Under different performance states, the reference value C^{Δ} of the maximum inter-story drift is exhibited in Table 5.

Table 5. Reference value C^{Δ} of θ_{max} under different performance states.

Performance States	Immediate Occupancy (IO)	Life Safe (LS)	Collapse Prevention (CP)
θ_{max}	1/100	1/50	1/25

The distribution diagrams of the maximum inter-story drift are obtained by amplitude modulation, as shown in Figure 13. The frame numbers of x and y represent earthquakes in the x and y direction, respectively. The maximum inter-story drift of each storey increased with increasing the peak ground acceleration. The maximum inter-story drift of frame KJ1 decreased as the storeys increased. Additionally, the maximum inter-story drift occurred in the first storey under different peak ground accelerations, which applied to the soft storey of the frame structures. Frame KJ2 adopted the enlarging cross-section method and set cross braces in the x direction; the maximum inter-story drift moved upwards, and the soft storey was transferred to the second or the third storey. However, the soft storey was still the first storey in the y direction. The maximum inter-story drift of frame KJ1 satisfied the IO state at 0.2 g; the LS state at 0.4 g; and the CP state near 0.8 g. The maximum inter-story drift of frame KJ2 satisfied the IO state at 0.4 g; the LS state near 0.8 g; and the CP state near 1.8 g. Besides, compared to frame KJ1, the maximum inter-story drift of frame KJ2 was significantly reduced under the same the peak ground acceleration. Additionally, the effect of the enlarging cross-section method and setting the cross braces was more significant.

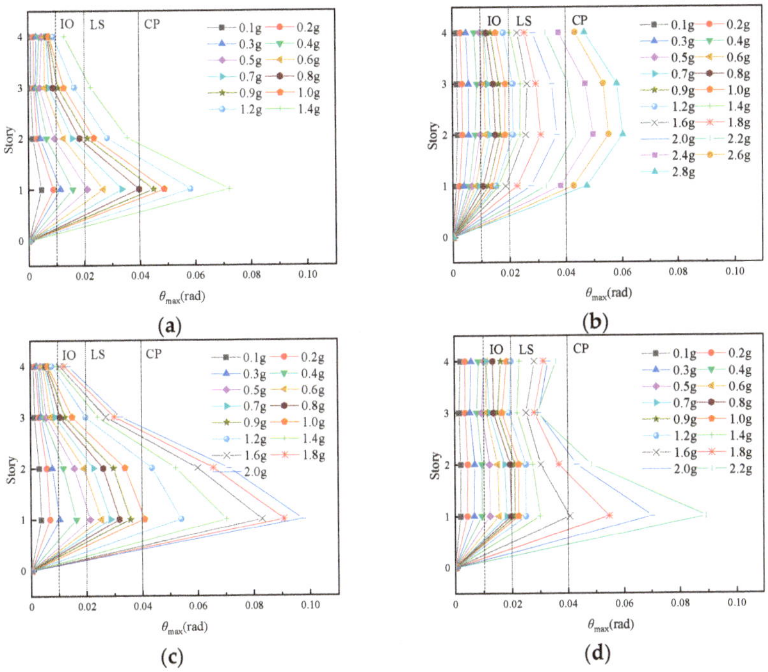

Figure 13. Distribution diagram of maximum inter-story drifts. (**a**) KJ1-x; (**b**) KJ2-x; (**c**) KJ1-y; (**d**) KJ2-y.

5.2. Distribution Diagram of Residual Inter-Story Drifts

FEMA 356 [30] defines four performance points for the residual inter-story drift (θ_R) and gives the reference value C^{Δ} for θ_R in different performance states. In this paper, three performance points were selected for analysis, as shown in Table 6. Among them, DS1 corresponded to the minor repair state; DS2 corresponded to the major repair state; and DS3 corresponded to the collapse state.

Table 6. Reference value C^{Δ} of θ_R under different performance states.

Performance States	DS1	DS2	DS3
θ_R	1/500	1/200	1/100

The distribution diagrams of the residual inter-story drift are obtained via amplitude modulation, as plotted in Figure 14. The residual inter-story drift of the frame structures decreased with the increase in storeys, and the maximum residual inter-story drift occurred on the first storey under different peak ground accelerations. The residual inter-story drift of the frame structures increased with the increase in the peak ground acceleration. Frame KJ1 reached the DS1 state at 0.4–0.5 g, DS2 state at 0.5–0.7 g, and DS3 state at 0.7–0.9 g; Additionally, frame KJ2 reached the DS1 state at 1.8 g under the x-direction earthquake; however, frame KJ2 reached the DS1 state at 1.0 g, DS2 state at 1.6 g, and DS3 state at 1.8 g under the influence of the y direction earthquake. Compared with frame KJ1, the residual inter-story drift of frame KJ2 was significantly reduced, and the seismic performance and seismic resilience were significantly improved under the same peak ground acceleration.

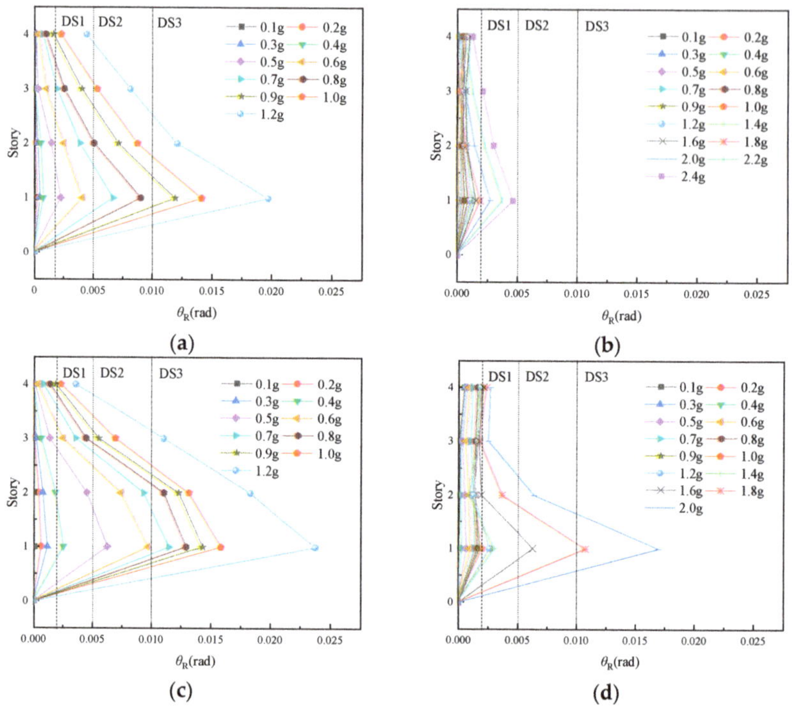

Figure 14. Distribution diagram of residual inter-story drifts. (**a**) KJ1-x; (**b**) KJ2-x; (**c**) KJ1-y; (**d**) KJ2-y.

5.3. Incremental Dynamic Analysis

Under the action of the No. 1 earthquake ground motion record, the incremental dynamic analysis of the frame structures was carried out before and after strengthening, as illustrated in Figure 15. When a_{PG} was small, the initial stage of the IDA curve showed approximately linear growth, indicating that the structures were in an elastic state. As a_{PG} increased, the slope of the IDA curve gradually decreased, indicating that the structures were in an elastic–plastic state.

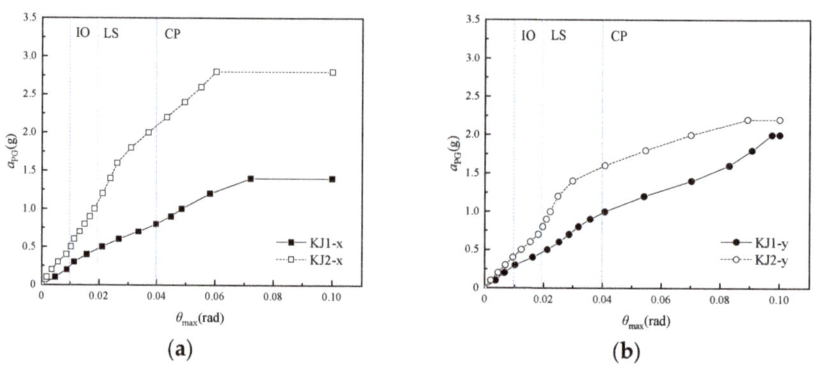

Figure 15. IDA curve diagram of the No. 1 earthquake ground motion record. (**a**) x-direction earthquakes (**b**) y-direction earthquakes.

IDA curves and their summary fractile curves are depicted in Figure 16 under the action of twenty earthquake ground motion records. Statistical analysis was conducted according to θ_{max}, and the median and logarithmic standard deviation of θ_{max} were calculated. Thus, fractile curves of 16%, 50%, and 84% were obtained. It could be seen that (1) the shape of IDA curves was different. The curves had characteristics of "excessive softening", "excessive hardening", and "fluctuation". The IDA curve could comprehensively reflect the possible seismic responses of frame structures under different intensity levels. (2) Under the same earthquake direction, the difference between the mean values of θ_{max} was significant for frames KJ1 and KJ2. The IDA curve of frame KJ2 showed overall data points moving to the left, which indicated that the seismic performance of the existing frame structure could be significantly improved by the enlarging cross-section method and the external self-centering substructure.

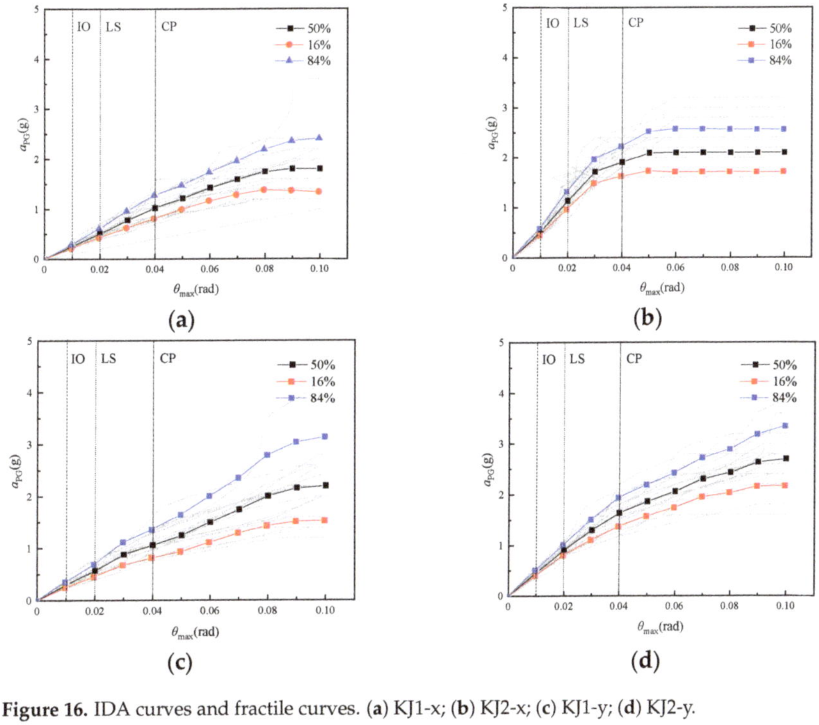

Figure 16. IDA curves and fractile curves. (**a**) KJ1-x; (**b**) KJ2-x; (**c**) KJ1-y; (**d**) KJ2-y.

6. Seismic Fragility Analysis of RC Frame Structures

6.1. Probabilistic Seismic Demand Models

The median value of the seismic demand capacity $\overline{D}$ and the peak ground acceleration a_{PG} follow an exponential relationship [31], as shown in Equation (1):

$$\overline{D} = \alpha a_{PG}^{\beta} \tag{1}$$

Both sides of the equation take the natural logarithm at the same time. Linear regression was conducted using mathematical software. Additionally, the probabilistic seismic demand model curve of frame structures was obtained (as displayed in Figure 17), as given in Equation (2):

$$\ln \overline{D} = \ln\alpha + \beta \ln a_{PG} \tag{2}$$

where lnα and β are the coefficient of statistical regression after a large number of incremental dynamic analyses.

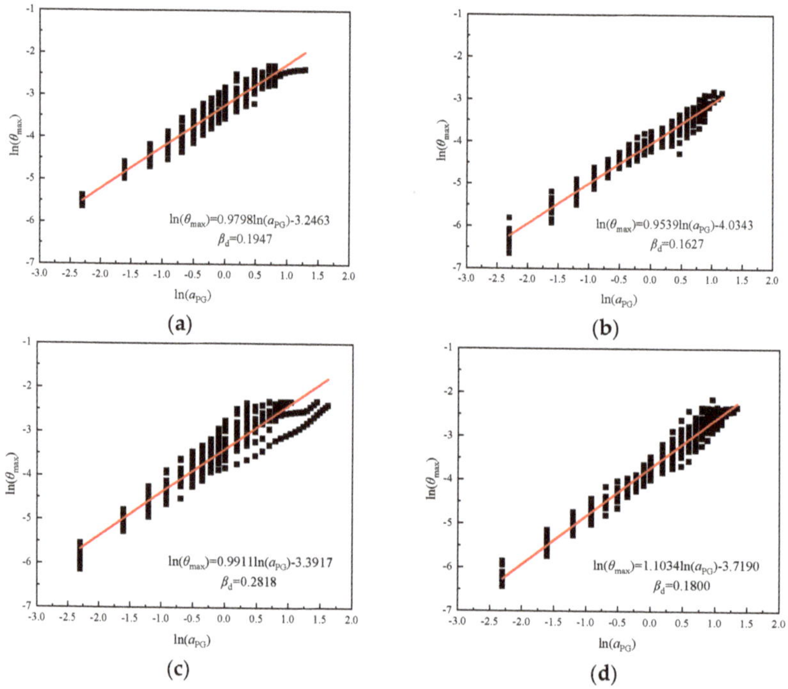

Figure 17. Regression curve of probabilistic seismic demand models. (**a**) KJ1-x; (**b**) KJ2-x; (**c**) KJ1-y; (**d**) KJ2-y.

The standard deviation β_d of the logarithmic normal distribution function for the seismic demand capacity is presented in Equation (3):

$$\beta_d = \sqrt{\frac{\sum_{i=1}^{N} [\ln D_i - \ln \overline{D}]}{N-2}} \tag{3}$$

where N is the number of data points for the regression analysis, and D_i is the result of the time history analysis (i = 1, 2, . . ., N).

It could be seen from Figure 17 that the curve of the probabilistic seismic demand model was approximately linear, the logarithmic standard deviation of the seismic demand parameter β_d was approximately 0.2, and the fitting effect was good.

6.2. Seismic Fragility Curve of Frame Structures

After obtaining the probabilistic seismic demand model, the total probability method is used to calculate the exceedance probability of the frame structures under each performance state. The formula is given in Equation (4) [32]:

$$P_f = \Phi\left[\frac{\ln\left[\alpha(PGA)^{\beta}/C^{\Delta}\right]}{\sqrt{\beta_c^2+\beta_d^2}}\right] \tag{4}$$

where β_c is the logarithmic standard deviation of the seismic capacity of the frame structures, and the uncertainty was not considered.

The seismic fragility curves of the frame structures are plotted in Figure 18. It is apparent from Figure 18 that the exceedance probability of each performance level truly reflects the seismic performance level of frame structures under different peak ground accelerations. (1) The development law of the seismic fragility curve is similar, and the overall curve is S-shaped; (2) when the peak ground acceleration is relatively low, the frame structure is approximately intact, or undergoes slight damage, and the fragility curves basically overlap under different performance states; (3) as the peak ground acceleration increases, the elastic–plastic deformation of the frame structures fully develops, and the slope of the curve firstly increases and then decreases. The performance level of the frame structures gradually develops from immediate occupancy to collapse; (4) compared with frame KJ1, the exceedance probability of frame KJ2 decreases significantly at various performance levels under the same peak ground acceleration, especially under the action of the earthquake in the x direction. The exceedance probability of frame KJ2 in the IO state and LS state is even lower than that of frame KJ1 in the LS state and CP state.

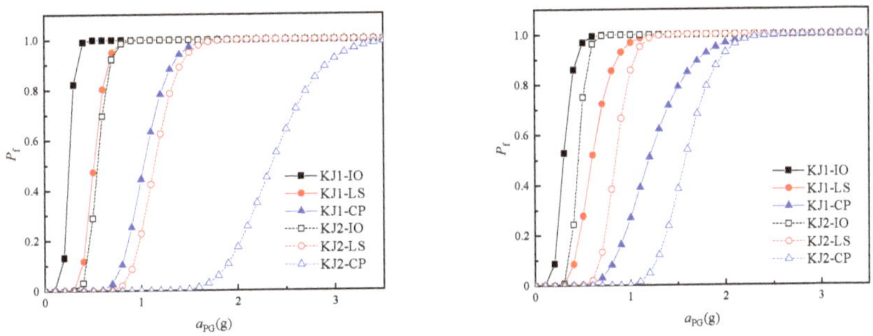

(**a**) (**b**)

Figure 18. Seismic fragility curve. (**a**) x-direction earthquakes; (**b**) y-direction earthquakes.

A comparison of failure states is illustrated in Figure 19. As an example of the analysis, the failure state of the frame structures is compared under the action of the earthquake in the x direction. Comparing Figure 19a,b, it can be seen that when a_{PG} is 0.2 g, the exceedance probability of the IO state for frame KJ2 increases from 86.9% to 100.0%. When a_{PG} is 0.4 g, the exceedance probability of the IO state increases from 0.9% to 96.9%. When a_{PG} is 0.6 g, the exceedance probability of the IO state increases from 0.0% to 30.4%; the exceedance probability of the LS state increases from 19.8% to 69.6%; and the exceedance probability of the CP state decreases from 79.9% to 0.0%. When a_{PG} is 1.0 g, the exceedance probability of the LS state increases from 0.0% to 77.4%; the exceedance probability of the CP state decreases from 55.6% to 22.6%; and the exceedance probability of the collapse state decreases from 44.4% to 0.0%. When a_{PG} is 2.0 g, the exceedance probability of the CP state increases from 0.0% to 82.8%, and the exceedance probability of the collapse state decreases from 100.0% to 17.2%. It is evident that the exceedance probability of the frame structures is significantly reduced, and the safety margin is greatly increased.

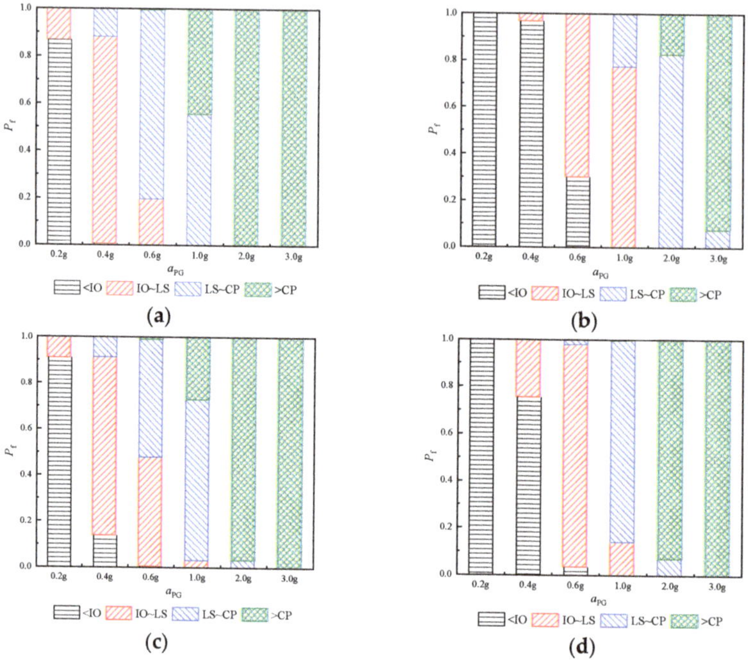

Figure 19. Comparison of failure states. (**a**) KJ1-x; (**b**) KJ2-x; (**c**) KJ1-y; (**d**) KJ2-y.

7. Conclusions

Finite element models of one-story one-span frame structures and four-storey frame structures were established using OpenSees finite element platform. The seismic response laws of the frame structures were analyzed under twenty earthquake ground motion records. Additionally, the seismic fragility analysis of the frame structures was conducted. The main conclusions are as follows:

(1) The proposed modeling method was reasonable, and the error between the experiment results and finite element simulation results was small;
(2) Under the same peak ground acceleration, the maximum inter-story drift of frame KJ1 decreased with the increase in storeys. However, the maximum inter-story drift of frame KJ2 moved upwards, and was significantly reduced;
(3) When the peak ground acceleration was small, the IDA curves showed a linear increase. As the peak ground acceleration increased, the slope of the IDA curves gradually decreased;
(4) Based on the results of IDA, the probabilistic seismic demand model curve was fitted, and the fitting effect was good;
(5) When the same peak ground acceleration was applied to the frame structures, the exceedance probability of frame KJ2 was significantly lower than that of frame KJ1 at various performance levels.

Author Contributions: Methodology, H.L., M.H.; software, F.W., J.L.; writing—original draft preparation, W.L.; writing—review and editing, J.Z.; supervision, J.Z. All authors have read and agreed to the published version of the manuscript.

Funding: This research was funded by [the National Natural Science Foundation of China] grant number [52178448].

Data Availability Statement: The data that has been used is confidential.

Conflicts of Interest: The authors declare no conflict of interest.

References

1. *GB 50011-2010*; Code for Seismic Design of Buildings. China Architecture and Building Press: Beijing, China, 2016. (In Chinese)
2. Heydari, P.; Mostofinejad, D.; Mostafaei, H.; Ahmadi, H. Strengthening of deep RC coupling beams with FRP composites: A numerical study. *Structures* **2023**, *51*, 435–454. [CrossRef]
3. Ortiz, J.D.; Khedmatgozar Dolati, S.S.; Malla, P.; Nanni, A.; Mehrabi, A. FRP-reinforced/strengthened concrete: State-of-the-art review on durability and mechanical effects. *Materials* **2023**, *16*, 1990. [CrossRef]
4. Zhou, C.Y.; Wang, L.N.; Wang, Y.; Fang, Z. Experimental study on the flexural strengthening of one-way RC slabs with end-buckled and/or externally bonded CFRP sheets. *Eng. Struct.* **2023**, *282*, 115832. [CrossRef]
5. Li, J.B.; Gong, J.X.; Wang, L.H. Experimental study on seismic behavior of corroded reinforced concrete columns strengthened with concrete jacket and CFRP. *China Civ. Eng. J.* **2009**, *42*, 17–26. (In Chinese) [CrossRef]
6. Huang, J.F.; Zhu, C.M.; Gong, Z.G.; Zhang, F.W. Experimental study on seismic behavior of earthquake-damaged RC frame strengthened by enlarging cross-section. *China Civ. Eng. J.* **2012**, *45*, 9–17. (In Chinese)
7. Song, L.L.; Guo, T.; Gu, Y.; Cao, Z.-L. Experimental study of a self-centering prestressed concrete frame subassembly. *Eng. Struct.* **2015**, *88*, 176–188. [CrossRef]
8. Priestley, M.H.N. Overview of PRESSS research program. *PCI J.* **1991**, *36*, 50–67. [CrossRef]
9. Priestley, M.H.N. The PRESSS program current status and proposed plans for phase III. *PCI J.* **1996**, *41*, 22–40. [CrossRef]
10. Lu, X.L.; Cui, Y.; Liu, J.J.; Gao, W. Shaking table test and numerical simulation of a 1/2-scale self-centering reinforced concrete frame. *Earthq. Eng. Struct. Dyn.* **2015**, *44*, 1899–1917. [CrossRef]
11. Cui, Y.; Lu, X.L.; Jiang, C. Experimental investigation of tri-axial self-centering reinforced concrete frame structures through shaking table tests. *Eng. Struct.* **2017**, *132*, 684–694. [CrossRef]
12. Kurosawa, R.; Sakata, H.; Qu, Z.; Suyama, T. Precast prestressed concrete frames for seismically retrofitting existing RC frames. *Eng. Struct.* **2019**, *184*, 345–354. [CrossRef]
13. Eldin, M.N.; Dereje, A.J.; Kim, J. Seismic retrofit of framed buildings using self-centering PC frames. *J. Struct. Eng.* **2020**, *146*, 04020208. [CrossRef]
14. Huang, Z.H.; Cai, L.P.; Pandey, Y.; Tao, Y.; Telone, W. Hysteresis effect on earthquake risk assessment of moment resisting frame structures. *Eng. Struct.* **2021**, *242*, 112532. [CrossRef]
15. Ge, F.W.; Tong, M.N.; Zhao, Y.G. A structural demand model for seismic fragility analysis based on three-parameter lognormal distribution. *Soil Dyn. Earthq. Eng.* **2021**, *147*, 106770. [CrossRef]
16. Ji, J.; Elnashai, A.S.; Kuchma, D.A. An analytical framework for seismic fragility analysis of RC high-rise buildings. *Eng. Struct.* **2007**, *29*, 3197–3209. [CrossRef]
17. Sarno, L.D.; Pugliese, F. Seismic fragility of existing RC buildings with corroded bars under earthquake sequences. *Soil Dyn. Earthq. Eng.* **2020**, *134*, 106169. [CrossRef]
18. Sarno, L.D.; Pugliese, F. Effects of mainshock-aftershock sequences on fragility analysis of RC buildings with ageing. *Eng. Struct.* **2021**, *232*, 111837. [CrossRef]
19. Kumar, P.; Samanta, A. Seismic fragility assessment of existing reinforced concrete buildings in Patna, India. *Structures* **2020**, *27*, 54–69. [CrossRef]
20. Choudhury, T.; Kaushik, H.B. Treatment of uncertainties in seismic fragility assessment of RC frames with masonry infill walls. *Soil Dyn. Earthq. Eng.* **2019**, *126*, 105771. [CrossRef]
21. Pitilakis, D.; Petridis, C. Fragility curves for existing reinforced concrete buildings, including soil-structure interaction and site amplification effects. *Eng. Struct.* **2022**, *269*, 114733. [CrossRef]
22. Gautam, D.; Adhikari, R.; Rupakhety, R. Seismic fragility of structural and non-structural elements of Nepali RC buildings. *Eng. Struct.* **2021**, *232*, 111879. [CrossRef]
23. Palagala, V.Y.; Singhal, V. Structural score to quantify the vulnerability for quick seismic assessment of RC framed buildings in India. *Eng. Struct.* **2021**, *243*, 112659. [CrossRef]
24. Dalal, S.P.; Dalal, P. Strength, deformation and fragility assessment of reinforced concrete moment resisting frame designed by force based design and the performance based plastic design method for seismic loads. *Structures* **2021**, *29*, 1154–1164. [CrossRef]
25. Cao, X.Y.; Feng, D.C.; Wu, G. Seismic performance upgrade of RC frame buildings using precast bolt-connected steel-plate reinforced concrete frame-braces. *Eng. Struct.* **2019**, *195*, 382–399. [CrossRef]
26. Mander, J.B.; Priestley MJ, N.; Park, R. Theoretical stress-strain model for confined concrete. *J. Struct. Eng.* **1988**, *114*, 1804–1826. [CrossRef]
27. Liu, H.; Zeng, Y.X.; Han, M.J.; Fan, Y.L.; Zhang, G.W.; Qin, C.A. Experimental research of seismic behavior of existing RC frame strengthened with attached diagonal cable brace. *Build. Struct.* **2023**. (In Chinese) [CrossRef]
28. Fan, Y.L.; Liu, H.; Han, M.J.; Zeng, Y.X.; Zhang, G.W.; Qin, C.A. Study on seismic performance of self-centering member reinforced frame based on finite element. *Archit. Technol.* **2023**, *54*, 1174–1177. (In Chinese)
29. Vamvatsikos, D.; Cornell, C.A. Incremental dynamic analysis. *Earthq. Eng. Struct. Dyn.* **2002**, *31*, 491–514. [CrossRef]
30. American Society of Civil Engineers; Federal Emergency Management Agency. *Prestandard and Commentary for the Seismic Rehabilitation of Buildings*; Federal Emergency Management Agency: Washington, DC, USA, 2000.

31. Cornell, C.A.; Jalayer, F.; Hamburger, R.O.; Foutch, D.A. Probabilistic basis for 2000 SAC Federal Emergency Management Agency steel moment frame guidelines. *J. Struct. Eng.* **2002**, *128*, 526–533. [CrossRef]
32. Yu, X.H. *Probabilistic Seismic Fragility and Risk Analysis of Reinforced Concrete Frame Structures*; Harbin Institute of Technology: Harbin, China, 2012. (In Chinese)

buildings

MDPI

Article

A Study on the Applicability and Accuracy of the Discrete Element Method for Plates Based on Parameter Sensitivity Analysis

Fei Guo [1,2] and Jihong Ye [1,2,*]

1 Jiangsu Key Laboratory Environmental Impact and Structural Safety in Engineering, China University of Mining and Technology, Xuzhou 221116, China; gflykey@126.com
2 Xuzhou Key Laboratory for Fire Safety of Engineering Structures, China University of Mining and Technology, Xuzhou 221116, China
* Correspondence: jhye@cumt.edu.cn

Abstract: In order to verify the accuracy and applicability of the discrete element method (DEM) in dealing with geometrically large deformations of continuous plate structures, both a single-parameter analysis and an orthogonal design method were adopted to analyze the displacement responses of the plate structures and were compared with those calculated using the finite element method (FEM). The single-parameter change condition involved the thickness-to-width ratio, elastic modulus, or Poisson's ratio, while the multi-parameter change included boundary conditions, dimensions, load forms, thickness-to-width ratio, elastic modulus, and Poisson's ratio. The results showed that displacements of the target locations were basically identical to those obtained according to FEM, with a maximum error of less than 5% under the single-parameter change condition. The maximum displacement error of the plate structures calculated using the DEM and FEM, respectively, was 4.212%, and the mean error and extreme difference of error parameters were 2.633% and 2.184%, respectively. These results indicate that the displacements of the plate structures calculated using the DEM were highly consistent with those obtained according to the FEM. Additionally, single-parameter changes and multi-parameter changes barely influenced the accuracy and suitability of the DEM in solving displacement response problems of plate structures. Therefore, the DEM is applicable in terms of dealing with displacement response problems of plate structures.

Keywords: discrete element method; plate structures; parameter sensitivity analysis; orthogonal design method; error analysis

Citation: Guo, F.; Ye, J. A Study on the Applicability and Accuracy of the Discrete Element Method for Plates Based on Parameter Sensitivity Analysis. *Buildings* **2023**, *13*, 1567. https://doi.org/10.3390/buildings13061567

Academic Editor: Elena Lucchi

Received: 11 May 2023
Revised: 2 June 2023
Accepted: 14 June 2023
Published: 20 June 2023

1. Introduction

As a common type of structure in daily life, the plate structure is widely used in engineering, aerospace, ships, water conservancy, and other fields, and investigations on the corresponding structural deformations under various loads are of great significance. The commonly used plate calculation theories include the classical thin plate theory based on elastic surface differential equations [1], the thin plate theory based on Kirchhoff's hypothesis of straight normal lines [2], and the Reissner–Mindlin moderately thick plate theory based on the consideration of transverse shear deformation, the method of which needs to construct interpolation functions and is highly required in terms of element continuity [3,4]. For the geometrically nonlinear problem of plates, the effect of membrane stress on the plate surface needs to be considered. The control equations and deformation coordination equations are both complex high-order differential equation systems, which are difficult to solve. Therefore, power series solutions and trigonometric series solutions are usually used for simple calculations of plates, during the process of which nonlinear equation systems are solved at a slow convergence speed [5–7]. In engineering practice, due to the complexity of the load and boundary conditions of the plate structure, it is difficult to obtain accurate results by using analytical methods. Numerical methods are mostly used for analysis, including meshless methods [8], boundary element methods [9], and finite element methods [10]. Among them, the finite element method is widely used, and scholars have proposed various computational formats, such as the total Lagrangian method (TL) and the updated Lagrangian method (UL) [11–17]. With the development of computer technology, the finite element method has become the most commonly used numerical analysis method in the field of engineering structures, and its calculation accuracy and efficiency are widely recognized. In the finite element method, the structural deformations are reflected by node displacements, the solution process of which needs to construct continuous shape functions and the calculation domain and displacement field need to be kept continuous because inappropriate displacement patterns may make it difficult to accurately simulate the mechanical behaviors of the structures. For example, a single element type is often not applicable for solving problems involving plates with different thicknesses, and certain low-order elements are usually inaccurate in terms of simulating the bending deformations of the plate structures due to the phenomenon of "shear locking". For geometrically nonlinear problems, it is necessary to invert the structural stiffness matrix and repeat and iterative solutions, often encountering computational inefficiencies due to non-convergence.

The traditional discrete element method (DEM) has been developed for more than 40 years, initially proposed as a numerical calculation method for solving problems in granular mechanics such as rocks and soils [18]. The method discretizes an object into a collection of rigid particles and simulates the object under certain kinds of loads by tracking the positions of the particles according to the force–displacement relationship between adjacent particles and the kinematic equations. Timsina and Christy proposed the Applied Element Method (AEM) to analyze the fracture and collapse of reinforced concrete [19] and masonry structures [20] under cyclic loads. LE et al. [21] used the discrete element method to simulate damage and crack propagation in composite plates, eventually realizing simulations of fiber delamination and fracturing. KUMAR et al. [22,23] studied the influence of the microstructural characteristics of discrete particles on the macroscopic behavior of structures and used the discrete element method to simulate the buckling of compressed cylinders. Professor Ye Jihong's research group [24–29] proposed a DEM model for truss structures and derived an expression of the spring contact stiffness coefficient for truss DEM, which was also applied to the vibration, buckling, large deformation, and elastoplastic analyses of frame and grid shell structures. To address the mechanical responses of plate structures under different loading conditions, Guo [30] proposed a single-layer particle arrangement plate discrete element method and derived the contact spring stiffness coefficient between particles based on the principle of energy conservation. Displacement continuity and deformation compatibility were not required by adopting the above-mentioned method, thereby avoiding convergence issues. Additionally, the discrete particles were purely rigid and did not undergo any deformations themselves, and their rotation and translation were independent of each other. Therefore, the DEM was applicable for solving problems involving plates with varying thicknesses, and there was no such phenomenon of "shear locking".

Sensitivity analysis methods are commonly used to determine the influence degree of different parameters on target performance, including single-parameter sensitivity analysis methods [31,32] and multi-parameter sensitivity analysis methods [33,34], which respectively consider the influence of a single-parameter variation and the coupling of multiple parameter variations on the control target performance. Cavaliere [35] employed a commercial multi-objective optimization tool to precisely determine the weight of each parameter on the reduction behavior and found that temperature was the main factor influencing the time to total reduction. Batou [36] studied a new model updating method, which updated each model parameter separately by constructing a measurement output transformation that was only sensitive to itself for each parameter. Zhang M [37] proposed an algebraic and direct method for solving the sensitivity of complex modal parameters of asymmetric systems and verified its effectiveness and correctness through single-parameter numerical experiments and multi-parameter numerical experiments. Wang et al. [38] examined the effects of hydrogen relative humidity, air relative humidity, operating temperature, and the air stoichiometry ratio on the performance of polymer electrolyte membrane fuel cells through orthogonal experimental design.

Reference [30] provided stiffness coefficients of contact elements in the single-layer particle arrangement plate discrete element method and developed a plate discrete element calculation program in Fortran. However, the generality of this algorithm has not been validated. In this investigation, sensitivity analysis methods are adopted to analyze the influence of different parameter variations on the accuracy and applicability of plate deformations based on the discrete element method and deformations of the plate structures based on the DEM were also compared with those calculated according to the finite element method. The influence of parameter variations on the accuracy of a plate-deformation-based discrete element method is analyzed, providing a basis for further verifying the universality of the algorithm.

2. Basic Theory of the DEM for a Plate

2.1. The Model of the DEM for a Plate

The model of a square plate based on DEM is created by discretizing the plate into a row of spherical particles, which can be divided into corner particles, edge particles, and interior particles according to the corresponding positions, as illustrated in Figure 1. It should be noted that determinations of the particle radius are related to the sizes of the plate surface rather than plate thickness.

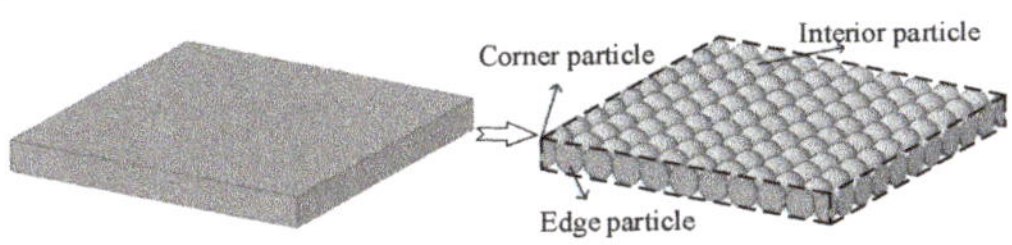

Figure 1. DEM model of the plate structure.

In the plate discrete element model, adjacent particles are connected by artificially defined zero-length springs to form the basic analysis unit of the discrete plate element, namely the contact element. According to the location of the contact element, the contact element can be divided into edge contact, interior contact, and diagonal contact. The diagonal contact element only contains one normal spring, while other contacts contain six springs, including one normal, two tangential, one torsion, and two bending springs, as shown in Figure 2.

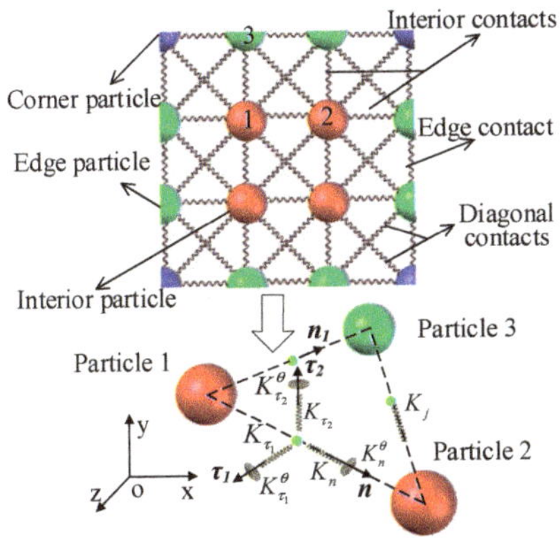

Figure 2. Contact element.

All particle movements follow Newton's second law. Take a discrete particle α as an example, and assume that it is connected with other n units, and the external force and external moment acting on particle α are $\mathbf{F}^{ext}$ and $\mathbf{M}^{ext}$, respectively. According to Newton's second law, the equation of motion for particle α can be expressed as:

$$\begin{cases} \mathbf{m}_\alpha \frac{d^2\mathbf{r}}{dt^2} = \sum_{j=1}^{n} \mathbf{F}_j^{\text{int}} + \mathbf{F}^{ext} \\ \mathbf{J}_\alpha \frac{d\omega}{dt} = \sum_{j=1}^{n} \mathbf{M}_j^{\text{int}} + \mathbf{M}^{ext} \end{cases} \tag{1}$$

In the equation, $\mathbf{m}_\alpha$ and $\mathbf{J}_\alpha$ represent the mass and inertia moment of particle α, respectively. $\mathbf{r}$ and $\boldsymbol{\omega}$ represent the displacement vector and angular velocity vector of particle α. $\mathbf{F}_j^{\text{int}}$ and $\mathbf{M}_j^{\text{int}}$ represent the contact force and contact torque generated by the j-th unit adjacent to particle α, and t represents time.

2.2. Calculation of Internal Forces in Contact Elements

It can be obtained from Figure 2 that the material properties and deformation of the contact element are closely related to the zero-length spring of the contact point between the two particles. The stiffness of each spring along different directions is independent and does not affect each other. By combining Equation (1) with the central difference method, the relative linear displacement and relative angular displacement between the two particles in the contact element can be obtained. By adopting Equation (2), the incremental internal force at the contact point can also be calculated. According to the principle of force translation, the contact force is transferred to the center of the particle and integrated to obtain the internal force at the center of the particle.

$$\Delta\mathbf{S} = \mathbf{K}_e \Delta\mathbf{U} \tag{2}$$

In the equation, $\Delta\mathbf{S}$ represents the incremental internal force of the contact element, including internal force and internal moment; $\Delta\mathbf{U}$ represents the incremental relative displacement between the two particles of the contact element under the local coordinate system, including relative linear displacement and angular displacement. $\mathbf{K}_e$ represents the elastic stiffness matrix of the contact element, which includes three translational and three rotational stiffness for both edge and internal contacts, and only one normal stiffness for diagonal contacts, and the corresponding values can be calculated through the following equations [30].

$$Edge\ contact: \begin{cases} K_n' = \frac{Eh}{2(1+\mu)} \\ K_{\tau_1}' = \frac{(1-3\mu)Eh}{8(1-\mu^2)} \\ K_{\tau_2}' = \frac{Eh}{4(1+\mu)} \\ K_n^{\theta\prime} = \frac{Eh^3}{24(1+\mu)} \\ K_{\tau_1}^{\theta\prime} = \frac{Eh^3}{24(1-\mu^2)} \\ K_{\tau_2}^{\theta\prime} = \frac{ER^2h}{6(1-\mu^2)} \end{cases} \tag{3}$$

$$Interior\ contact: \begin{cases} K_n = \frac{Eh}{1+\mu} \\ K_{\tau_1} = \frac{(1-3\mu)Eh}{4(1-\mu^2)} \\ K_{\tau_2} = \frac{Eh}{2(1+\mu)} \\ K_n^{\theta} = \frac{Eh^3}{12(1+\mu)} \\ K_{\tau_1}^{\theta} = \frac{Eh^3}{12(1-\mu^2)} \\ K_{\tau_2}^{\theta} = \frac{ER^2h}{3(1-\mu^2)} \end{cases} \tag{4}$$

$$Diagonal\ contact : K_j = \frac{\mu E h}{1-\mu^2} \tag{5}$$

In the equation, E, h, μ, and R represent the elastic modulus, thickness, Poisson's ratio, and particle radius of the plate, respectively.

3. Theory of Parameter Sensitivity Analysis

When using the discrete element method to analyze the deformation characteristics of plate structures, uncertainties in model input may lead to uncertainties in the accuracy of output results. The sensitivity analysis method investigates and analyzes the influence of various input uncertainties on the output uncertainties of the model [39], including single-parameter sensitivity analysis and multi-parameter sensitivity analysis.

3.1. Single-Parameter Sensitivity Analysis

Single-parameter sensitivity analysis is a method to investigate the influences of changes in a single parameter on the performance of the control objective within a certain range while keeping other parameters constant. It is also known as perturbation analysis. Single-parameter sensitivity analysis has been widely used in the fields of structure and materials due to easy operation and intuitive understanding.

3.2. Multi-Parameter Sensitivity Analysis

Multi-parameter sensitivity analysis considers the interaction between parameters, resulting in more reasonable and scientific results. Common methods for multi-parameter sensitivity analysis include the full factorial method and orthogonal design method.

3.2.1. Full Factorial Method

The key to the full factorial method is comparisons and analyses of all design combinations under different operating levels with various influential factors, which are able to provide a large amount of data for parameter analysis to accurately evaluate the interaction between parameter factors. It is commonly used for multi-parameter analysis when there are few parameters and levels to be considered for parameter interaction. Full factorial design requires a combination of $n_1 \times n_2 \ldots \times n_i \ldots \times n_j$, where n_i represents the number of levels for the i-th factor, and j represents the number of design parameters. Figure 3a shows a three-parameter, three-level full factorial design with a total of 27 combination sample points.

Figure 3. The method of sensitivity analysis: (**a**) full factor design; (**b**) orthogonal design.

3.2.2. Orthogonal Design Method

If there are too many model parameters and the parameters have multiple operating levels, using a full factorial design will result in a large number of operating combinations, which leads to a huge computational burden. The orthogonal design method selects representative parameters from each parameter and operating level to combine and judges the interaction effects between parameters and the corresponding influences on control objectives. Figure 3b shows a three-parameter, three-level orthogonal design with a total of nine combination sample points. Therefore, the orthogonal design method is an efficient design method for arranging multiple parameter combinations scientifically and is characterized by uniform dispersion, regularity, and comparability. The orthogonal design method can significantly reduce the number of analysis samples while considering the interaction of multiple parameters. Figure 4 shows the form and code meanings of orthogonal tables, and Table 1 is a four-parameter, three-level orthogonal table, which shows that the number of occurrences of each parameter operating level in any column is the same, and the arrangement of numbers in any two columns is complete and balanced.

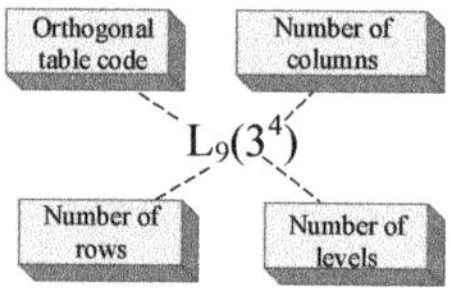

Figure 4. Orthogonal table form and code meaning.

Table 1. Orthogonal representation.

Operating Combination	Parameter			
	A	B	C	D
1	1	1	1	1
2	1	2	2	2
3	1	3	3	3
4	2	1	2	3
5	2	2	3	1
6	2	3	1	2
7	3	1	3	2
8	3	2	1	3
9	3	3	2	1

4. Parameter Sensitivity Analysis

In this paper, a rectangular plate is used as the fundamental numerical model, as illustrated in Figure 5. The discrete element method was employed to calculate the deformations of the plate under varying parameter conditions and was compared with those obtained from the finite element method to validate the accuracy of the algorithm based on DEM. The finite element method considers the effect of in-plane membrane stress by enabling large deflection. The discrete element method uses particle centroid displacement as the basic quantity since the displacements obtained by solving the motion equation itself are reflections of force responses of the structure. Similarly, the finite element method also reflects the structure deformations through the nodal displacements. Therefore, using displacement as a reference for comparison between the two methods is more convenient and intuitive.

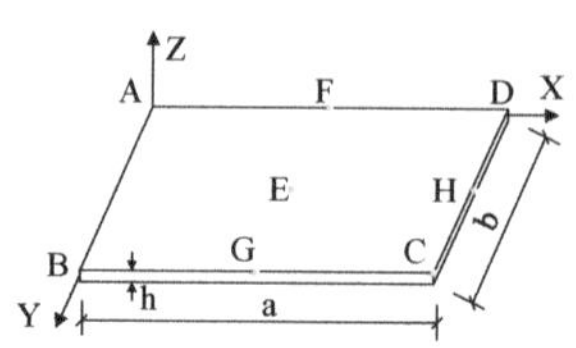

Figure 5. Rectangle plate model.

4.1. *Influences of Single-Parameter Changes*

When considering the influence of a single-parameter change, the plate thickness-to-width ratio, elastic modulus, and Poisson's ratio were respectively applied to Equations (3)–(5) as basic analysis parameters to investigate the influences of single-parameter changes on the accuracy of the algorithm. For the model shown in Figure 5, the values of a and b were set to be 0.3 m, and ϱ was set to be 7850 kg/m^3. The boundary conditions of the simulation model included a four-edge simple support and a four-edge fixed support. A uniform load q was applied to the plate by using a static loading method. There were 21 particles and 420 contact elements in the discrete element model, as shown in Figure 6. The radius of the spherical particles was 15 mm. The time step, Δt, was set to be 1×10^{-6} s, and the total calculation time was 0.5 s.

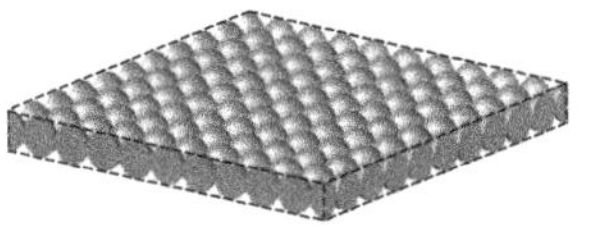

Figure 6. DEM model of the square plate.

4.1.1. Thickness-to-Width Ratio

In dealing with deformation problems of loaded plates by using the finite element method, plates were classified based on their thickness-to-width ratio into the following categories: (1) thick plates: $1/8 < h/b < 1/5$; (2) thin plates: $1/80 < h/b < 1/8$; and (3) membrane: $1/100 < h/b < 1/80$. In this investigation, when solid186 elements were employed to simulate thick plates based on FEM, the shear effect along the thickness direction of the plate was considered. When Shell181 elements were used for thin plates and membranes, the in-plane stress of the membrane was considered to avoid the phenomenon of "shear locking" in the plate. The selection of finite element analysis elements for plates of different thicknesses in the subsequent examples followed this rule. For the displacement response problem of plates with different thicknesses and loads, no special treatment is needed when using the discrete element method. For four-edge simply supported and four-edge fixed boundary conditions, the uniformly distributed load, q, on the plate was set to be 9.6×10^6 Pa and 9.6×10^7 Pa, respectively, with the elastic modulus $E = 2.1 \times 10^{11}$ Pa and Poisson's ratio $\mu = 0.24$. According to the range of plate thickness, the thickness-to-width ratio h/b was set to be 0.01, 0.015, 0.02, 0.04, 0.06, 0.08, 0.1, 0.12, 0.14, 0.16, 0.18, and 0.2, respectively. The discrete element method was used to calculate the deflection-to-span ratio at the center point of the plate with different thickness-to-width ratios, and the results were compared with those obtained based on FEM.

According to Figure 7, it is evident that plates with smaller thicknesses exhibited larger deflections. Bending deformations of the plate were obvious, and the in-plane membrane stress could not be ignored. As the plate thickness increased, shear deformations along the thickness direction became significant. The results obtained from the displacement response analyses of plates with different thicknesses, as shown in Figures 7 and 8, using the discrete element calculation were in good agreement with those obtained using the finite element calculation. With increasing plate thickness, structural deformations reduced, and the calculation error slightly increased but remained within 5%. Compared to the finite element method, the discrete element method for plates did not require changing the element type for plates of different thicknesses, and it minimized the occurrence of "shear locking". Therefore, within the range of plate properties, the thickness did not significantly affect the accuracy and applicability of the discrete element algorithm for plates.

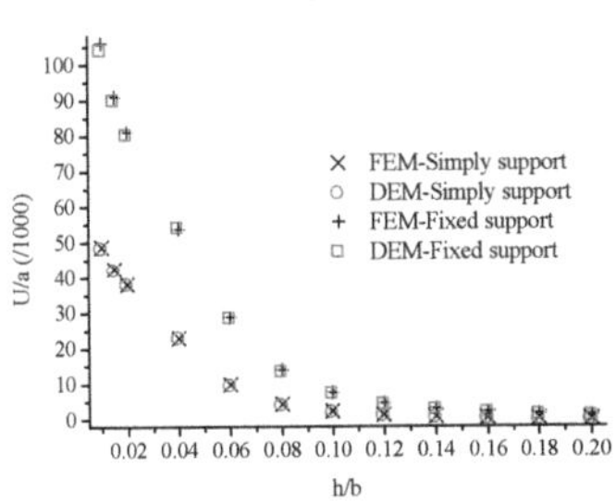

Figure 7. The effect of thickness-to-width ratio variation on the deflection-to-span ratio of the plate.

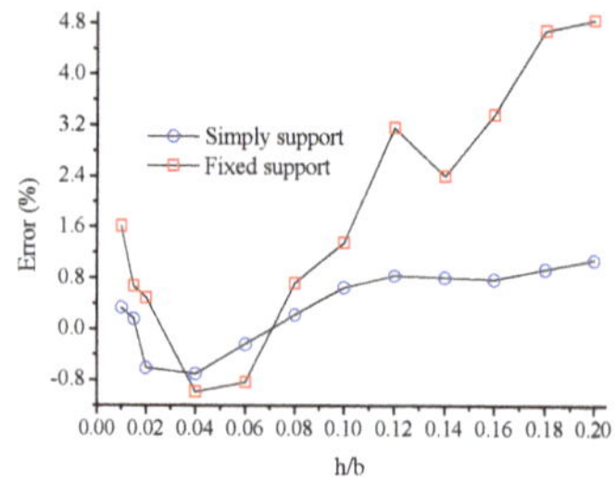

Figure 8. Influence of different thickness-to-width ratios on DEM–FEM error.

4.1.2. Elastic Modulus

For square plates with the boundary conditions of four-edge simple support and four-edge fixed support, the uniformly distributed loads applied on the plate were set to be 9.6×10^6 Pa and q = 9.6×10^7 Pa, respectively. The plate thickness, h, was set to be 6 mm and the Poisson's ratio, μ, was set to be 0.24. The basic elastic modulus of the plate, E_0, under both boundary conditions, was set to be 2.1×10^{11} Pa, and the range of the elastic modulus varies from 0.1 E_0 to 10.0 E_0 with a decrement of 0.5 E_0, the variation process of which included small and large deformations of the plate and a total of 21 analysis scenarios. The discrete element method was used to calculate changes in the deflection-span ratio of the center point of the square plate with different elastic moduli, and the corresponding results were compared with those obtained according to the finite element method.

It can be observed from Figure 8 that the deflections of the plate were large with decreasing elastic modulus, the bending deformation effect was significant, and the stress of the in-plane membrane cannot be ignored. It can also be observed from Figures 9 and 10 that both algorithms considered the effect of in-plane membrane stress and the results obtained based on the discrete element method and the finite element method were basically consistent with a maximum DEM–FEM error of 3.54%, indicating that the variation of elastic modulus had little effect on the accuracy and applicability of the DEM algorithm.

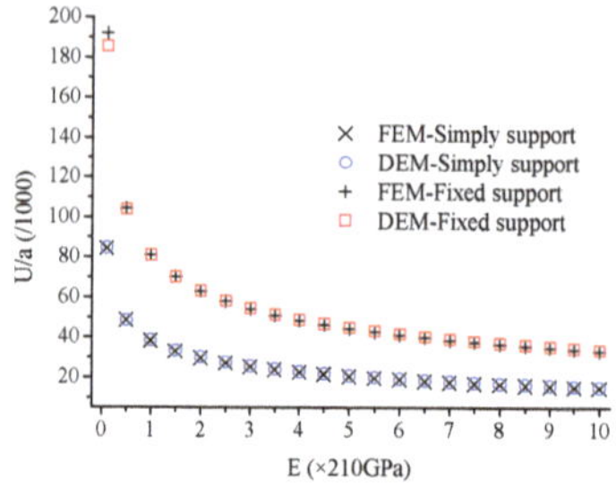

Figure 9. Change in the deflection-to-span ratio of the plate center point under different elastic moduli.

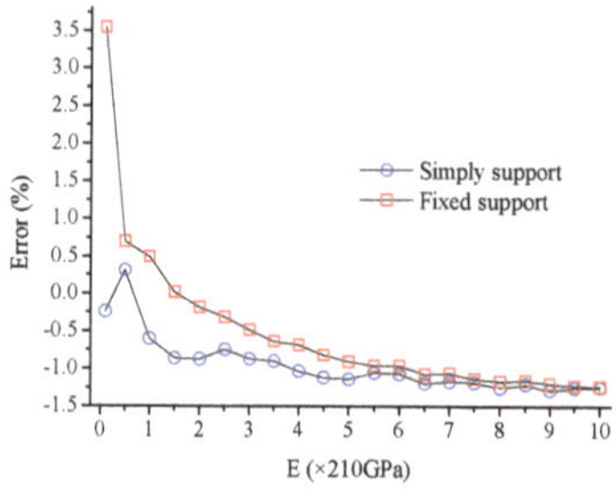

Figure 10. Influence of elastic modulus variation on DEM–FEM error.

4.1.3. Poisson's Ratio

As a material property of the plate itself, Poisson's ratio also affects its own bending stiffness under unchanged external loads. For square plates with a boundary condition of four-edge simple support and four-edge fixed support, the uniformly distributed loads, q, applied on the plate were set to be 9.6×10^6 Pa and 9.6×10^7 Pa, respectively. Plate thickness, h, was set to be 6 mm, and elastic modulus, E, was set to be 2.1×10^{11} Pa. Poisson's ratio ranged from 0.1 to 0.48, with a decrement of 0.02, covering the Poisson's ratio range of commonly used materials. A total of 20 analysis scenarios were considered. The discrete element method was used to calculate the changes in the deflection–span ratio of the center point of the thin plate only when Poisson's ratio changed, and the corresponding results were compared with those obtained according to the finite element method.

Changes in Poisson's ratio alter the bending stiffness of the plate, thereby affecting the deflection degrees of the plate under loads. As shown in Figures 11 and 12, as Poisson's ratio increased, deflections at the center of the plate slightly decreased. The results obtained from the discrete element calculation and finite element calculation were essentially consistent, with a maximum error of −2.11%, which indicated that changes in Poisson's ratio had little effect on the accuracy and applicability of the DEM algorithm.

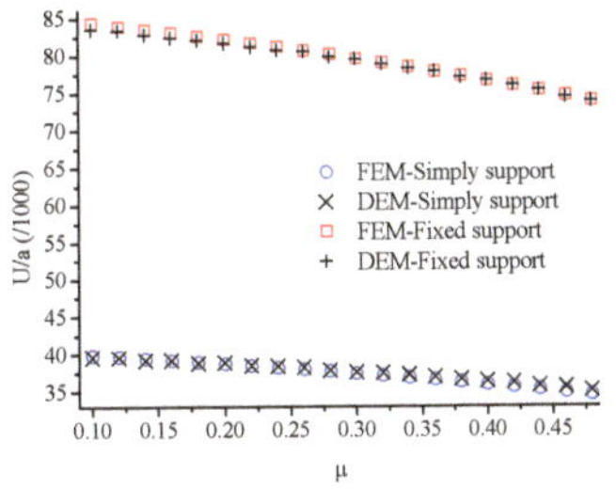

Figure 11. Change in the deflection-to-span ratio of the plate center point under different Poisson's ratios.

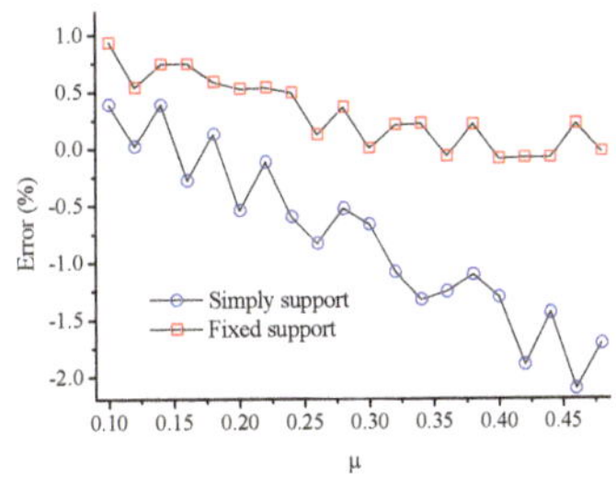

Figure 12. Influence of Poisson's ratio variation on DEM–FEM error.

4.2. Influences of Multi-Parameter Changes

In addition to the thickness-to-width ratio, elastic modulus, and Poisson's ratio, differences in plate boundary conditions, size effects, and load forms may also affect the applicability of the DEM algorithm. To further verify the universality of the DEM algorithm, the orthogonal design method was adopted to analyze the influences of multi-parameter changes on the accuracy and applicability of the DEM algorithm. The values of each parameter were determined as follows.

4.2.1. Boundary Conditions

Common boundary conditions of a rectangular plate include one-sided, two-sided, three-sided, and four-sided constraints, which can be classified as simple support and fixed support. Determinations of boundary conditions are not only related to the number of constrained sides but also the distributions of fixed and simple supports. To confirm the

DEM algorithm's applicability to any boundary condition, a mixed boundary form was applied that considers the influences of various factors, as presented in Table 2.

Table 2. Information about different boundary conditions.

Boundary	Fixed	Simply	Displacement Tracking Point	Position of Line Load	Position of Point Load
Cantilever plate	AB	\	C	CD	C
Two adjacent edges	AB	AD	C	CD	C
Two opposite edges	AB	CD	E	FG	E
Three edges	AB	AD, BC	H	CD	H
Four edges	AB, CD	BC, AD	E	FG	E

4.2.2. Plate Dimensions

There is a wide range of plate sizes in engineering, spanning from small plate components to large plate materials. According to the relevant literature, the plate lengths of the corresponding simulation model varied from 0.3 m to 10 m. In this investigation, five representative and diverse plate sizes (a × b) were chosen: 0.3 m × 0.3 m, 1.2 m × 0.6 m, 2 m × 2 m, 4 m × 1 m, and 10 m × 10 m, which considered both plate size effect and aspect ratios.

4.2.3. Form of Loading

Five common types of loads in engineering were selected for analyses, including uniformly distributed loads, concentrated loads, line loads, impact loads, and harmonic loads. Impact loads and harmonic loads were uniformly distributed on the entire plate surface, and the time-displacement curves of DEM–FEM were obtained by comparing the results obtained based on DEM and FEM transient analysis. The other load forms were analyzed by using static loading, and the load-deflection ratio curves of DEM–FEM were obtained by comparing the results obtained according to DEM and FEM static analysis. The position of the line load and concentrated load and the displacement tracking points are shown in Table 2, and the loading method is shown in Figure 13. To obtain obvious deformations of the plate, load values were increased as much as possible while ensuring the convergence condition of the finite element analysis.

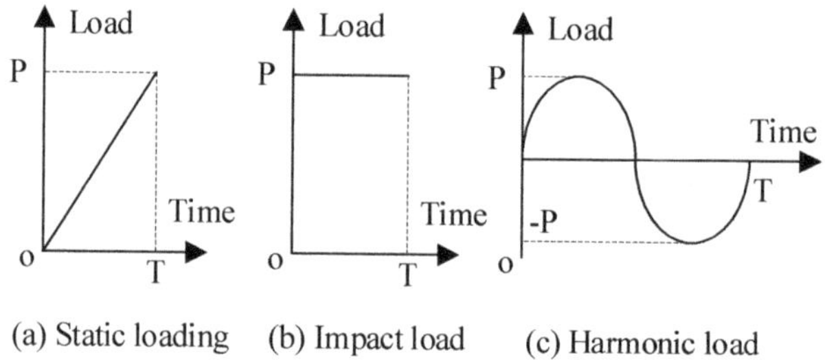

Figure 13. Loading mode.

4.2.4. Thickness-to-Width Ratio

Plates with a thickness-to-width ratio in the range of 0.125–0.2 are considered thick plates, while those with a thickness-to-width ratio in the range of 0.0125–0.125 are classified as thin plates, and those with a thickness-to-width ratio in the range of 0.01–0.0125 are regarded as membranes. To ensure the broad representativeness of parameter values, the thickness-to-width ratio factors of 0.01, 0.04, 0.1, 0.16, and 0.2 were used. Details about the computation units used for finite element analysis can be found in Section 4.1.1.

4.2.5. Elastic Modulus

The properties of common materials used in engineering are listed in Table 3. It shows that the elastic modulus of commonly used plates is less than 200 GPa. To ensure the broad representativeness of parameter values, five levels of elastic modulus were used, with values of 0.005 GPa, 50 GPa, 100 GPa, 150 GPa, and 200 GPa.

Table 3. Properties of commonly used materials.

Material	E (GPa)	μ	Material	E (GPa)	μ
Alloy steel	206	0.25–0.3	Cast steel	175	0.31–0.34
Lead	170	0.42	Aluminum alloy	71	0.3
Rolled aluminum	69	0.32–0.36	Concrete	14–23	0.1–0.18
Nylon	2.83	0.4	Rubber	0.00784	0.48

4.2.6. Poisson's Ratio

Table 4 shows that the Poisson's ratios of the commonly used materials in engineering are typically between 0.1 and 0.48. To ensure the generality and representativeness of the parameter values, five levels of Poisson's ratio were chosen, which are 0.1, 0.24, 0.3, 0.4, and 0.48.

Table 4. Analyzed parameters and levels.

	Parameters	Level 1	Level 2	Level 3	Level 4	Level 5
1	Boundary conditions	Cantilever plate	Two adjacent	Two opposite	Three edges	Four edges
2	Plate dimensions (m)	0.3 × 0.3	1.2 × 0.6	2 × 2	4 × 1	10 × 10
3	Form of loading	Uniformly distributed	Line load	Concentrated load	Impact	Harmonic
4	Thickness-to-width ratio	0.01	0.04	0.1	0.16	0.2
5	Elastic modulus (GPa)	0.005	50	100	150	200
6	Poisson's ratio	0.1	0.24	0.3	0.4	0.48

In this study, the orthogonal design method was utilized to comprehensively evaluate the effect of multi-parameter changes on the accuracy and applicability of the DEM algorithm. The levels of each parameter are summarized in Table 4. The orthogonal design method was performed on six parameters, where each parameter had five levels to reduce the number of analyzed cases from 15,625(5^6) cases obtained from the full-factorial method to 25(5^2) combinations obtained from the orthogonal design method. This significantly reduced the computational workload and enabled the interaction between parameters to be considered. The representativeness of the orthogonal design lay in the following aspects: (1) each column contained all levels of the respective parameter with equal frequency; (2) all possible combinations of any two columns occurred; and (3) due to the orthogonality of the orthogonal table, the test combinations were evenly distributed in the full-factorial design combinations. The orthogonal design combinations are shown in Table 5. The discrete element analysis results of each operating condition were compared with the corresponding finite element analysis results to calculate the errors between the two algorithms when considering multi-parameter changes. The DEM–FEM error comparison rules were as follows: for harmonic load, the error was compared based on the deflection peak value of the two algorithms; for impact load, the error was compared based on the stable solutions of the two algorithms; for other loads, the error was compared based on the final results of the two algorithms.

Table 5. Orthogonal design table for $L_{25}(5^6)$.

Operating Combination	Boundary	Dimension	Form of Loading	h/b	E	μ	Error * Y (%)
1	1	1	1	1	1	1	0.239
2	2	2	2	2	2	2	0.918
3	3	3	3	3	3	3	1.063
4	4	4	4	4	4	4	−2.342
5	5	5	5	5	5	5	2.002
6	1	2	3	4	5	1	0.838
7	2	3	4	5	1	2	1.359
8	3	4	5	1	2	3	3.102
9	4	5	1	2	3	4	4.212
10	5	1	2	3	4	5	2.805
11	1	3	5	2	4	1	0.187
12	2	4	1	3	5	2	0.474
13	3	5	2	4	1	3	2.762
14	4	1	3	5	2	4	1.623
15	5	2	4	1	3	5	0.424
16	1	4	2	5	3	1	0.460
17	2	5	3	1	4	2	−2.884
18	3	1	4	2	5	3	−1.736
19	4	2	5	3	1	4	0.865
20	5	3	1	4	2	5	1.747
21	1	5	4	3	2	1	−2.652
22	2	1	5	4	3	2	1.538
23	3	2	1	5	4	3	0.546
24	4	3	2	1	5	4	4.121
25	5	4	3	2	1	5	2.559
Mean value1 ** (%)	0.875	1.811	1.642	2.154	2.154	0.875	
Mean value2 ** (%)	1.435	0.718	2.213	1.922	2.008	1.435	
Mean value3 ** (%)	1.842	0.980	1.794	1.572	1.540	1.842	
Mean value4 ** (%)	2.633	2.100	1.703	1.845	1.753	2.633	
Mean value5 ** (%)	1.907	2.903	1.539	1.198	1.834	1.907	
Range *** (%)	1.757	2.184	0.674	0.956	0.614	1.757	

* The calculation formula for error is as follows: Error = $(U_{FEM} - U_{DEM})/U_{FEM} \times 100\%$, where U represents the deflection value at the analysis scenario. ** The mean value is denoted by Q_{ij}, where i and j represent the level number and parameter number, respectively. Q_{ij} indicates the mean error of the j-th parameter at the i-th level under the orthogonal combination. For example, $Q_{11} = (|Y_1| + |Y_6| + |Y_{11}| + |Y_{16}| + |Y_{21}|)/5$, which reflects the influence degrees of parameter level changes on the target value. *** The range is calculated as Range = $\max\{Q_j\} - \min\{Q_j\}$, which provides an estimate of the magnitude of the target value fluctuation due to parameter changes.

Table 5 shows that compared to the finite element results, the discrete element method can maintain high accuracy even when considering multi-parameter changes, such as boundary conditions, dimensions, load forms, thickness-to-width ratio, elastic modulus, and Poisson's ratio. In addition, the discrete element method did not require a change in element type for plates with different thicknesses. The error of maximum and mean values for each parameter level were 4.212% and 2.633% respectively, indicating that the DEM provided precise solutions to the displacement response problem of a loaded plate. The maximum range of the mean value for each parameter level was 2.184%, demonstrating the stability of the DEM algorithm. The deformation history curves of displacement tracking points under different conditions are illustrated in Figure 14. The DEM accurately tracked the displacement response of rectangular plates under loads, with no significant impact on the accuracy and applicability of the algorithm when considering multi-parameter changes. These findings confirm the universality of the DEM algorithm in solving deformation problems of rectangular plates under various loads.

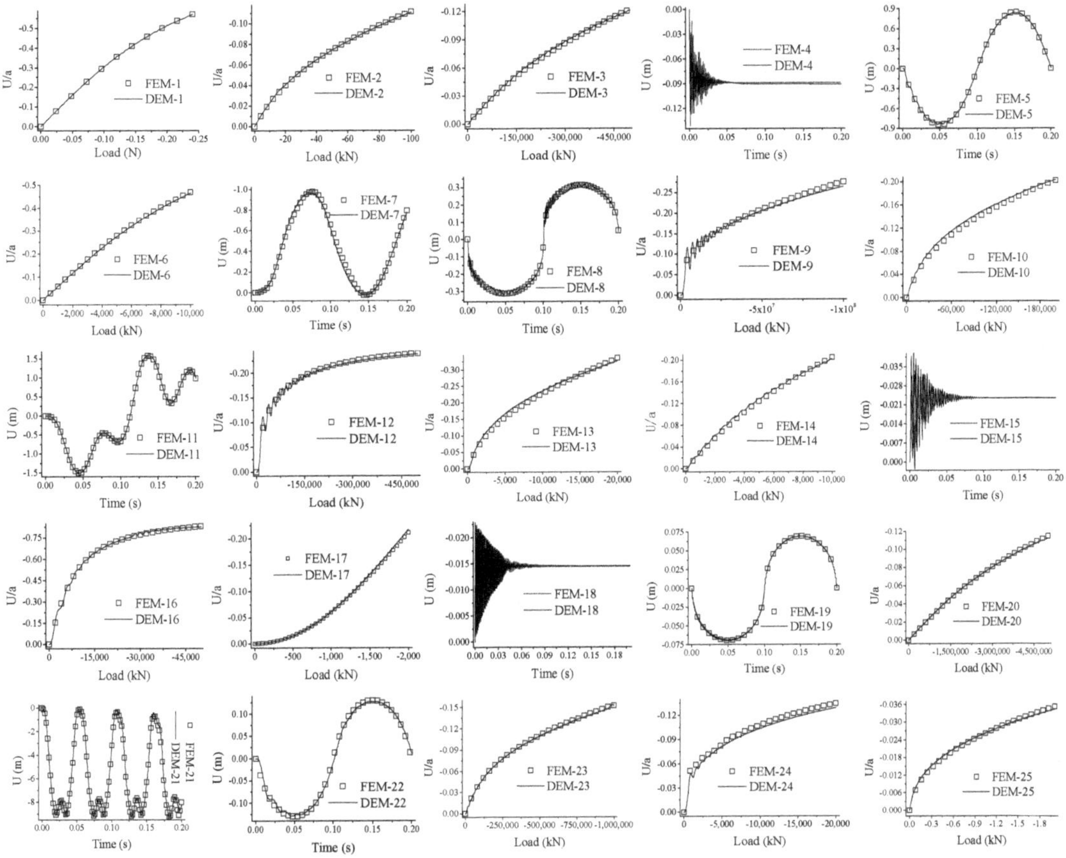

Figure 14. Deformation curve of the displacement tracking point under various working conditions.

5. Conclusions

In this investigation, both single-parameter analysis and orthogonal design methods were employed to analyze the impacts of relevant parameters on the displacement response of a rectangular plate under loading, utilizing the discrete element method. The following conclusions were drawn:

(1) The deformations of the plate were calculated and compared using the DEM and FEM through single-parameter analysis with varying thickness-to-width ratios, elastic modulus, or Poisson's ratios. The discrete element method for plates did not require a change in element type when the thickness-to-width ratio changed. With an increase in the thickness-to-width ratio, the deflection gradually decreased, and the FEM–DEM calculation error remained within 5%. Similarly, when the elastic modulus or Poisson's ratio changed, both methods considered the effect of membrane stress, with a maximum error of 3.54% and −2.11%, respectively. These results indicated that the variations in individual parameters had minimal impact on the accuracy and applicability of the DEM algorithm.

(2) To account for parameter interactions and their influence on the accuracy of the DEM algorithm, an orthogonal design method was adopted with five levels for six parameters: boundary conditions, dimensions, load forms, thickness-to-width ratio, elastic modulus, and Poisson's ratio. The displacement responses of the plate structure under loading were calculated for each combination and compared with the corresponding finite element results. Compared to the finite element method, the maximum error for the analysis scenarios was 4.212%. Furthermore, the mean error and extreme difference for each

parameter level were 2.633% and 2.184%, respectively. The static and dynamic displacement response curves of each working condition exhibited a high degree of consistency with those obtained according to the finite element method, further validating the universality of the plate discrete element method in addressing the force response problems in continuous medium plate structures.

Author Contributions: Conceptualization, methodology, software, validation, writing—original draft, writing, F.G.; supervision, funding acquisition, writing, J.Y. All authors have read and agreed to the published version of the manuscript.

Funding: This research was funded by the National Natural Science Foundation of China (Grant No. 51978655).

Data Availability Statement: Data sharing is not applicable to this article.

Conflicts of Interest: The authors declare no conflict of interest.

References

1. Bourada, M.; Bouadi, A.; Bousahla, A.A.; Senouci, A.; Bourada, F.; Tounsi, A.; Mahmoud, S.R. Buckling behavior of rectangular plates under uniaxial and biaxial compression. *Struct. Eng. Mech.* **2019**, *70*, 113–123.
2. Bohinc, U.; Brank, B.; Ibrahimbegovics, A. Discretization error for the Discrete Kirchhoff plate finite element approximation. *Comput. Methods Appl. Mech. Eng.* **2014**, *269*, 415–436. [CrossRef]
3. Nguyen-Xuan, H. A polygonal finite element method for plate analysis. *Comput. Struct.* **2017**, *188*, 45–62. [CrossRef]
4. Kikis, G.; Dornisch, W.; Klinkel, S. Adjusted approximation spaces for the treatment of transverse shear locking in isogeometric Reissner-Mindlin shell analysis. *Comput. Methods Appl. Mech. Eng.* **2019**, *354*, 850–870. [CrossRef]
5. Enshaeian, A.; Rofooei, F. Geometrically nonlinear rectangular simply supported plates subjected to a moving mass. *Acta Mech.* **2014**, *225*, 595–608. [CrossRef]
6. Kovalenko, M.D.; Abrukov, D.A.; Menshova, I.V.; Kerzhaev, A.P.; Yu, G.M. Exact solutions of boundary value problems in the theory of plate bending in a half-strip: Basics of the theory. *Z. Für Angew. Math. Und Phys.* **2019**, *70*, 98. [CrossRef]
7. Papkov, S.; Banerjee, J.R. A New Method for Free Vibration Analysis of Triangular Isotropic and Orthotropic Plates of Isosceles Type Using an Accurate Series Solution. *Mathematics* **2023**, *11*, 649. [CrossRef]
8. Belinha, J.; Aires, M. Elastoplastic Analysis of Plates with Radial Point Interpolation Meshless Methods. *Appl. Sci.* **2023**, *12*, 12842. [CrossRef]
9. Morse, L.; Mallardo, V.; Sharif-Khodaei, Z.; Aliabadi, F.M.H. Shape Optimisation of Assembled Plate Structures with the Boundary Element Method. *Aerospace* **2022**, *9*, 381. [CrossRef]
10. Burman, E.; Hansbo, P.; Larson, M.G. A simple approach for finite element simulation of reinforced plates. *Finite Elem. Anal. Des.* **2023**, *142*, 51–60. [CrossRef]
11. Trinh, M.C.; Jun, H. A higher-order quadrilateral shell finite element for geometrically nonlinear analysis. *Eur. J. Mech. A-Solids* **2021**, *89*, 104283. [CrossRef]
12. Lee, C.; Lee, D.H.; Lee, P.S. The strain-smoothed MITC3+ shell element in nonlinear analysis. *Comput. Struct.* **2022**, *265*, 106768. [CrossRef]
13. Wu, B.; Pagani, A.; Filippi, M.; Chen, W.Q.; Carrera, E. Large-deflection and post-buckling analyses of isotropic rectangular plates by Carrera Unified Formulation. *Int. J. Non-Linear Mech.* **2019**, *116*, 18–31. [CrossRef]
14. Rezaiee, P.M.; Arabie, E.; Masoodi, A.R. A triangular shell element for geometrically nonlinear analysis. *Acta Mech.* **2018**, *229*, 323–342. [CrossRef]
15. Ansarir, R.; Hasrati, E.; Shakouri, A.H. Nonlinear large deformation analysis of shells using the variational differential quadrature method based on the six-parameter shell theory. *Int. J. Non-Linear Mech.* **2018**, *106*, 130–143. [CrossRef]
16. Onkar, A.K. Nonlinear buckling analysis of damaged laminated composite plates. *J. Compos. Mater.* **2019**, *53*, 3111–3126. [CrossRef]
17. Ko, Y.; Leep, S.; Bathe, K.J. The MITC4+ shell element in geometric nonlinear analysis. *Comput. Struct.* **2017**, *185*, 1–14. [CrossRef]
18. Cundall, P.A.; Strack, O.D.L. A discrete numerical model for granular assemblies. *Geotechnique* **1979**, *29*, 47–65. [CrossRef]
19. Timsina, K.; Krishna, C.G.; Meguro, K. Sociotechnical Evaluation of the Soft Story Problem in Reinforced Concrete Frame Buildings in Nepal. *J. Perform. Constr. Facil.* **2021**, *35*, 04021019. [CrossRef]
20. Christy, D.L.; Pillai, T.M.M.; Nagarajan, P. Analysis of Brick Masonry Wall using Applied Element Method. In Proceedings of the International Conference on Recent Advances in Materials, Mechanical and Civil Engineering, Hyderabad, India, 1–2 June 2017.
21. Le, B.D.; Dau, F.; Charles, J.L. Modeling damages and cracks growth in composite with a 3D discrete element method. *Compos. Part B Eng.* **2016**, *91*, 615–630. [CrossRef]
22. Kumar, R.; Rommel, S.; Jauffres, D. Effect of packing characteristics on the discrete element simulation of elasticity and buckling. *Int. J. Mech. Sci.* **2016**, *110*, 14–21. [CrossRef]

23. Mathews, A.K.; Khan, A.; Sharma, B.; Kumar, S.; Kumar, R. A numerical investigation of granular shock waves over a circular cylinder using the discrete element method. *J. Fluid Mech.* **2022**, *936*, A11. [CrossRef]
24. Ye, J.H.; Qi, N. Progressive collapse simulation based on DEM for single-layer reticulated domes. *J. Constr. Steel Res.* **2017**, *128*, 721–731.
25. Ye, J.H.; Xu, L.L. Member Discrete Element Method for Static and Dynamic Responses Analysis of Steel Frames with Semi-Rigid Joints. *Appl. Sci.* **2017**, *7*, 714. [CrossRef]
26. Ye, J.H.; Zhang, M. Buckling behavior of single-layer reticulated shells based on member discrete element method. *J. Build. Struct.* **2019**, *40*, 50–57. (In Chinese)
27. Qi, N.; Ye, J.H. Nonlinear Dynamic Analysis of Space Frame Structures by Discrete Element Method. *Appl. Mech. Mater.* **2014**, *638–640*, 1716–1719. [CrossRef]
28. Xu, L.L.; Ye, J.H. DEM Algorithm for Progressive Collapse Simulation of Single-Layer Reticulated Domes under Multi-Support Excitation. *J. Earthq. Eng.* **2017**, *23*, 18–45. [CrossRef]
29. Xu, Q.; Ye, J.H. An adaptively coupled DEM–FEM algorithm for geometrical large deformation analysis of member structures. *Comput. Part. Mech.* **2019**, *7*, 947–959. [CrossRef]
30. Guo, F.; Ye, J.H. Contact Model and Elastic Deformation Analysis of Plate Structure Based on the Discrete Element Method. *Arch. Appl. Mech.* **2022**, *92*, 2513–2523. [CrossRef]
31. Borgonovo, E. A Methodology for Determining Interactions in Probabilistic Safety Assessment Models by Varying One Parameter at a Time. *Risk Anal.* **2010**, *30*, 385–399. [CrossRef]
32. Proppe, C. Local reliability based sensitivity analysis with the moving particles method. *Reliab. Eng. Syst. Saf.* **2021**, *207*, 107269. [CrossRef]
33. Frey, H.C.; Patil, S.R. Identification and review of sensitivity analysis methods. *Risk Anal.* **2010**, *22*, 553–578. [CrossRef]
34. Kamal, M.; Inel, M. Correlation between Ground Motion Parameters and Displacement Demands of Mid-Rise RC Buildings on Soft Soils Considering Soil-Structure-Interaction. *Buildigs* **2021**, *11*, 125. [CrossRef]
35. Cavaliere, P.; Perrone, A.; Marsano, D.; Primavera, V. Hydrogen-Based Direct Reduction of Iron Oxides Pellets Modeling. *Steel Res. Int.* **2023**, *94*, 2200791. [CrossRef]
36. Batou, A. A sensitivity-based one-parameter-at-a-time model updating method. *Mech. Syst. Signal Process.* **2019**, *122*, 247–255. [CrossRef]
37. Zhang, M.; Yu, L.; Xu, X. New Sensitivity Analysis Methods for Complex Modal Parameters in Asymmetrical Damped System. *AIAA J.* **2021**, *59*, 3165–3172. [CrossRef]
38. Wang, B.H.; Deng, C. Investigation of the effect of humidity at both electrode on the performance of PEMFC using orthogonal test method. *Int. J. Hydrogen Energy* **2019**, *44*, 13737–13743. [CrossRef]
39. Zhang, K.; Lu, Z.; Cheng, L. A new framework of variance based global sensitivity analysis for models with correlated inputs. *Struct. Saf.* **2015**, *55*, 1–9. [CrossRef]

buildings

MDPI

Article

Analysis of the Ultimate Load-Bearing Capacity of Steel-Clad Concrete-Filled Steel Tube Arched Protective Doors under Blast Shock Waves

Shangwei Dong, Zhimin Tian *, Xingwei Cao, Ce Tian and Zhenyu Wang

Institute of Defense Engineering, Academy of Military Sciences, People's Liberation Army, No. 24 Taiping Road, Beijing 100850, China

* Correspondence: overseatian@sina.com

Abstract: The mechanism of blast damage to steel-clad concrete-filled steel tube (SCCFST) arched protective doors is studied using the dynamic response characteristics of such loads under the action of blast shock wave loads, and the ultimate blast load-bearing capacity formula is derived based on the "plastic hinge" damage mode of the doors using limit analysis, which explores the effect of the blast shock wave. The effect of the design parameters of each component of the protective door on the load-bearing capacity subjected to blast shock waves is discussed. Results show that the damage mechanism under a uniform radial load on the outer surface of the SCCFST arched protective door is characterized by the plastic hinge lines at the two arch feet, which results in a slip fracture and renders the protective door unstable. The load-bearing capacity of the SCCFST arched protective door depends on the coordinated functioning of the cross-sectional outer cladding steel plate and inner connecting partition, concrete-filled steel tube, and restraining concrete outside the steel tube. The load-bearing capacity of each of the three parts differs with the varying cross-sectional occupancies.

Keywords: SCCFST arched protective door; dynamic response; damage mode; cross-sectional ultimate bending moment; ultimate bearing capacity

Citation: Dong, S.; Tian, Z.; Cao, X.; Tian, C.; Wang, Z. Analysis of the Ultimate Load-Bearing Capacity of Steel-Clad Concrete-Filled Steel Tube Arched Protective Doors under Blast Shock Waves. *Buildings* **2023**, *13*, 1424. https://doi.org/10.3390/buildings13061424

Academic Editors: Liqiang Jiang, Wei Chen, Chang He, Yi Hu, Qi Cai and Flavio Stochino

Received: 14 April 2023
Revised: 16 May 2023
Accepted: 24 May 2023
Published: 31 May 2023

1. Introduction

Protective doors are important building components used to ensure the safety of personnel and property inside a building. Understanding the blast damage mechanism of protective doors and determining their limited load-bearing capacity are critical to their design and application. The available literature indicates that flat and arched doors have been studied more extensively. In order to improve the blast resistance of protective doors, several researchers have conducted experimental studies and numerical analysis using software such as ANSYS and ABAQUS to optimize the structural system and form of the doors [1–4]. Meanwhile, novel door materials or filling materials are being used to improve door performance [5–11]. Compared to flat doors, arched doors offer better protection for the same span and weight, are better suited to withstand dynamic impact loads, and have significant advantages in blast resistance [12]. Steel-clad concrete-filled steel tube (SCCFST) structures exhibit excellent load-bearing performance, capitalizing on the advantages offered by steel and concrete materials. Additionally, these structures are easily processable. Hence, they can be widely applied to national defense and human defense engineering [13–15]. Considering these attributes, a satisfactory protective effect is expected to be achieved, and this study is a step in that direction.

Li et al. [16] carried out a dynamic characteristic analysis of a double steel-clad concrete-filled arched protective door and calculated the intrinsic frequencies and corresponding vibration patterns of the door under various boundary conditions while carrying out a spectral analysis of the overpressure time-history curve of the blast shock wave in front of the door. They found that the shock wave energy was concentrated in the low-frequency interval, covering the main vibration frequencies of the door. Chen et al. [17] investigated an arched panel with a double-layered structure and numerically studied its blast load resistance and energy absorption capacity using the finite element software LS-DYNA 971, showing that this arched panel structure had a higher blast load resistance than other forms of panels. Chen et al. [18] used AUTODYN to determine the blast load on a protective door in an arched reinforced concrete test. They compared the dynamic response characteristics of the door under single and multiple loads with the test, concluding that the dynamic response of the door was mainly influenced by the non-linear contact between the door and the frame and the strain rate effect. Guo et al. [19] used ABAQUS to study the dynamic response characteristics of SCCFST protective doors under the effect of explosions, and the analysis concluded that the steel pipe inside the doors contributed greatly to the enhancement of structural resistance, while the effect of increasing the wall thickness of the steel pipe and the strength of the concrete was not obvious.

Most studies use experimental and numerical simulation methods to analyze the damage phenomena and dynamic response laws of arched protective doors and steel-clad concrete structures, but the blast damage mechanism has been studied less. This essentially limits the application of the blast resistance-bearing capacity method in engineering design. To help develop engineering applications, we conducted numerical simulations to analyze the damage mechanism of SCCFST arched protective doors subjected to blast shock waves. Based on the motorized hinge method of limit analysis, a relationship was derived for the blast resistance-bearing capacity, and the influence of section design parameters on the load-bearing performance of the protective door was analyzed.

2. Problem Formulation and Computational Model

A typical two-side-supported arched protective door system is shown in Figure 1. The protective door consists of three components: concrete-filled steel tube, confined concrete, and surrounding steel panel and separators, as shown in Figure 2.

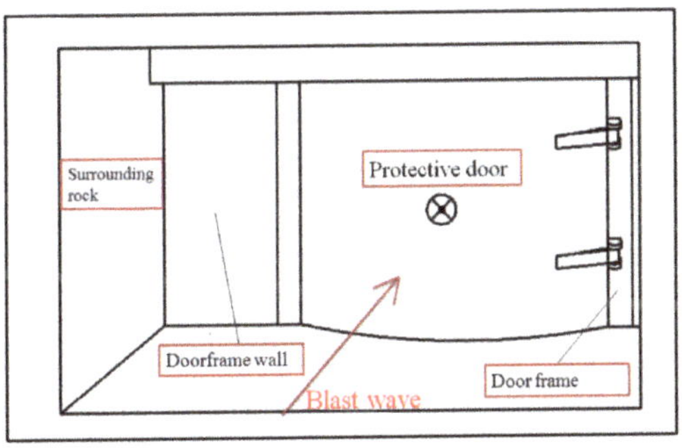

Figure 1. Schematic diagram of the arched protective door system.

As is shown in Figure 2b, the connection between the surface panel and its separators is realized by welding, as is the interface of the surface panel. Depending on the actual working conditions, either spot welding or welding through the length can be used. Supporting pads are used to isolate the steel tube and the surface panel, as well as the separators on the foot of the arch. After all the steel components are connected, concrete is poured inside.

In actual application, the net size of the aperture of the protective door is 3 m × 3 m. The parameters relevant to cross-section 1-1 are selected as follows: the inner diameter of the steel tube r_1 = 9.5 cm, the outer diameter r_2 = 11.5 cm, the minimum thickness of restrained concrete between the steel tube and the surrounding panel b_c = 1.0 cm, the separator thickness b_s = 0.5 cm, the net thickness of the concrete layer h_1 = 25 cm, the width of each cell b_1 = 25 cm, and the thickness of the surrounding panel h_s = 1.0 cm. The number of cells other than the edge cells is denoted as n, and in this study, n = 10. The arch axis radius R = 2.325 m and the circular angle 2θ = 90° are set for the cross-section 2-2.

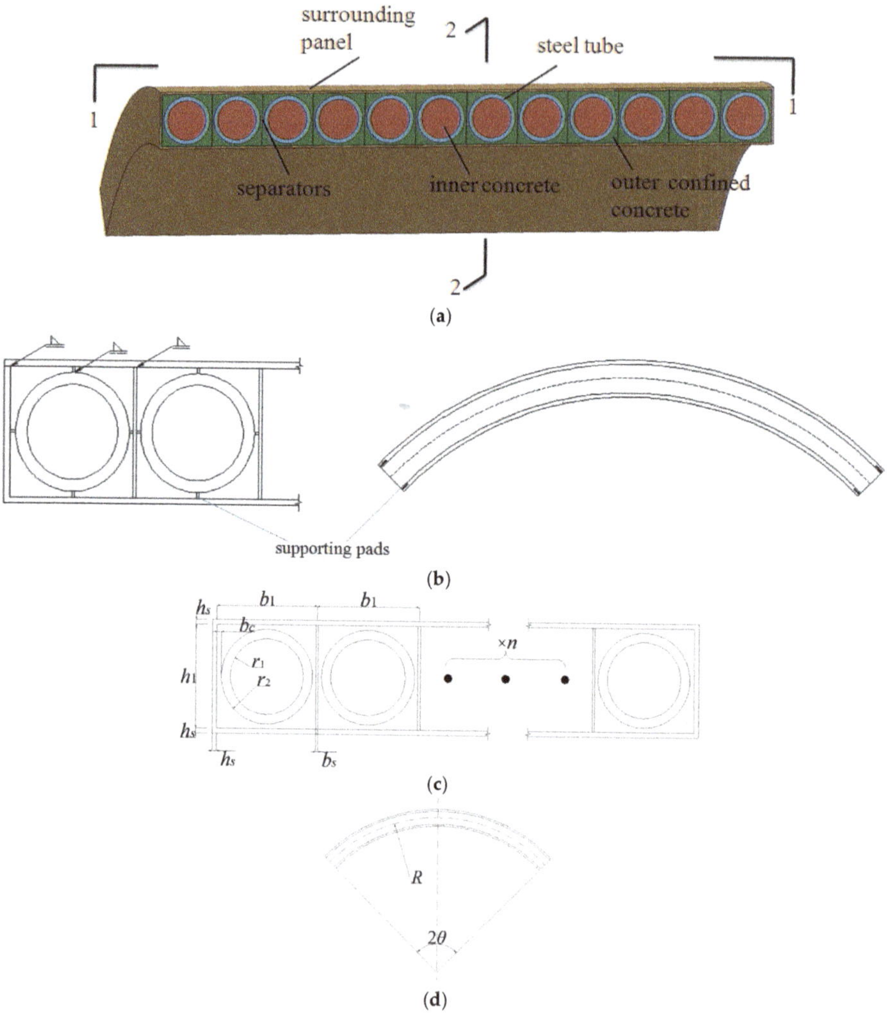

Figure 2. Schematic diagram of the SCCFST arched protective door. (**a**) Construction of the protective door. (**b**) Steel connections in the protective door. (**c**) Size of cross-section 1-1. (**d**) Size of cross-section 2-2.

The protective door system in Figure 1 can be simplified using the model shown in Figure 3. Based on the symmetry, a 1/2 model is built.

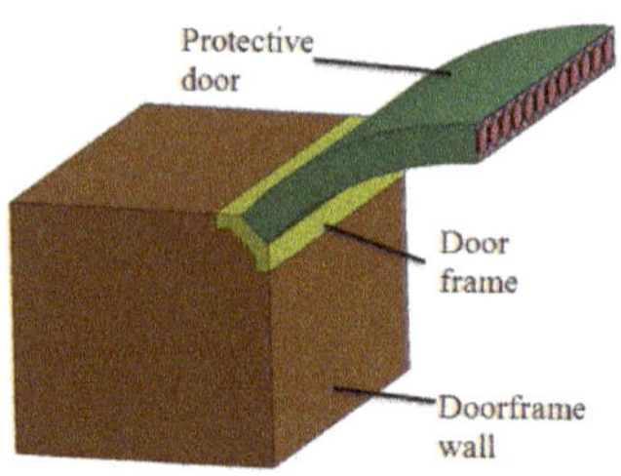

Figure 3. Schematic diagram of the numerical simulation model of the protective door.

The material properties are defined as follows: the steel is simulated using an intrinsic plastic strengthening model (Mat_Plastic_Kinematic) with yield strength f_y= 400 MPa, density = 7850 kg/m^3, and Young's modulus = 206 GPa. The concrete is simulated using the Johnson-Holmquist-Cook plastic damage model, with the design value of compressive strength being 27.5 MPa and that of tensile strength being 2.04 MPa. High-strength reinforced concrete is selected for the door frame wall to provide sufficient support for the protective door. The Mat_Concrete_Damage model is selected for simulation, with the axial compressive strength of matrix concrete f_c'= 50.2 MPa. The reinforcement is uniformly distributed in the matrix as equivalent steel content in the concrete. The steel content is taken as 0.5%, and the yield strength of the reinforcement f_y= 400 MPa [20].

Restraint is defined as the automatic contact between the protective door and door frame and the solid contact between the door frame and door frame wall, with a fixed-end restraint applied to the door frame wall.

The blast load is defined as a simplified sudden addition of a linear decay load without a boosting platform [16], as shown in Figure 4.

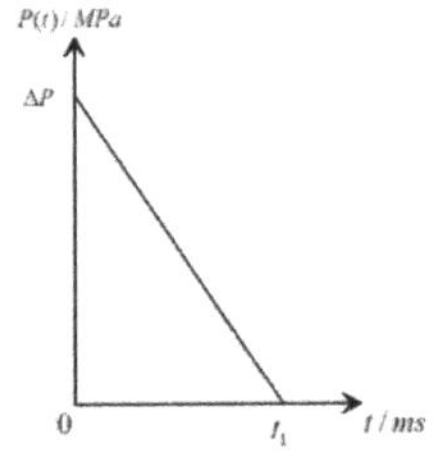

Figure 4. Blast load curve.

ΔP in Figure 4 is the peak overpressure of the blast load (MPa). Six calculating conditions with ΔP values of 7 MPa, 10 MPa, 11 MPa, 12 MPa, 13 MPa, and 14 MPa are selected, respectively. t_1 is uniformly taken as 100 ms. The load shown in Figure 4 is applied normally to the blast surface of the protective door and the door frame wall.

3. Dynamic Response Law and Damage Mode of the Protective Door under Blast Load

By analyzing the dynamic response law and damage mode of the protective door under different ΔP, the damage mechanism of the protective door subjected to a blast load provides a basis for theoretical calculation.

3.1. *Dynamic Response Law of Protective Door*

For the SCCFST arched protective door, the changes in the displacement of the arch top and 1/4 arch span position can represent the dynamic response law well [21]. Therefore, the arch top and 1/4 arch span positions are taken as the characteristic points for the displacement time-history analysis under different ΔP.

Figure 5 shows that at the top of the protective door arch, when ΔP is 7 MPa, the displacement time-history curve oscillates regularly about the zero scale, which shows the elastic dynamic response of the protective door. When ΔP is 10 MPa–12 MPa, the displacement time-history curve of the top of the arch is generally below the zero scale. After reaching the peak of negative displacement, it still presents regular damped vibration, which indicates the stable elastic-plastic vibration of the protective door. When ΔP reaches 13 MPa, there is a sudden change in the displacement time-history curve of the arch top, and there will no longer be stable vibration after reaching the peak of negative displacement. This indicates imminent plastic instability. Figure 6 shows that the displacement time-history change law on the 1/4 arch span is similar to that of the arch top, and there is a sudden change in the shape of the displacement time-history curve when ΔP reaches 13 MPa. This phenomenon coincides with the results observed by Budiansky and Roth in their study of the dynamic stability of spherical shells. They observed that, for a certain load level, when a small increment of the load causes a large change in the displacement response of the structure, the structure undergoes dynamic buckling, i.e., instability at the extreme point, and the corresponding load is referred to as the dynamic damage critical load [22]. This problem is also associated with a sudden change in the shape of the displacement time-history curve at the characteristic point when ΔP is between 12 MPa and 13 MPa. Then, the ΔP corresponding to the instant of sudden change can be considered as the ultimate load that the protective door can withstand.

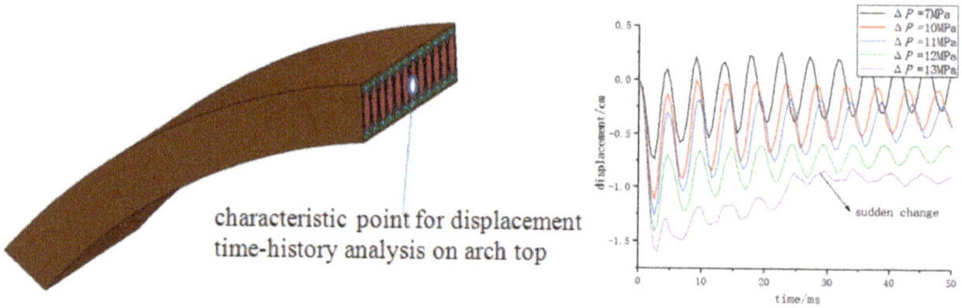

Figure 5. Displacement time-history on the top of the SCCFST arched protective door.

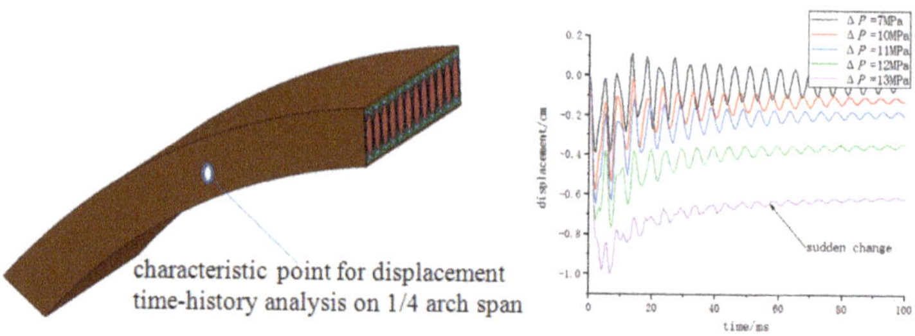

Figure 6. Displacement time-history on 1/4 arch span of the SCCFST arched protective door.

3.2. Damage Mode of the Protective Door

For observation and analysis, an element from the middle of the protective door is selected, and the damage phenomena and damage mode are analyzed for each component. See Figure 7.

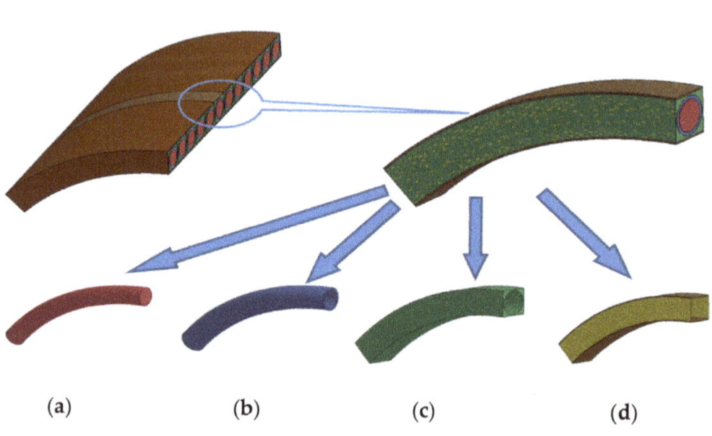

Figure 7. Schematic diagram of the components of a cell of the protective door. (**a**) Inner concrete (**b**) Steel tube (**c**) Confined concrete (**d**) Surface panel and its separators.

Two typical working conditions of ΔP= 10 MPa and ΔP = 12 MPa are selected, and the damage phenomena of each component in the protective door cell are presented in Figures 8 and 9:(In the figures, e refers to ×, e.g., 8e + 03 refers to 8 × 10^3).

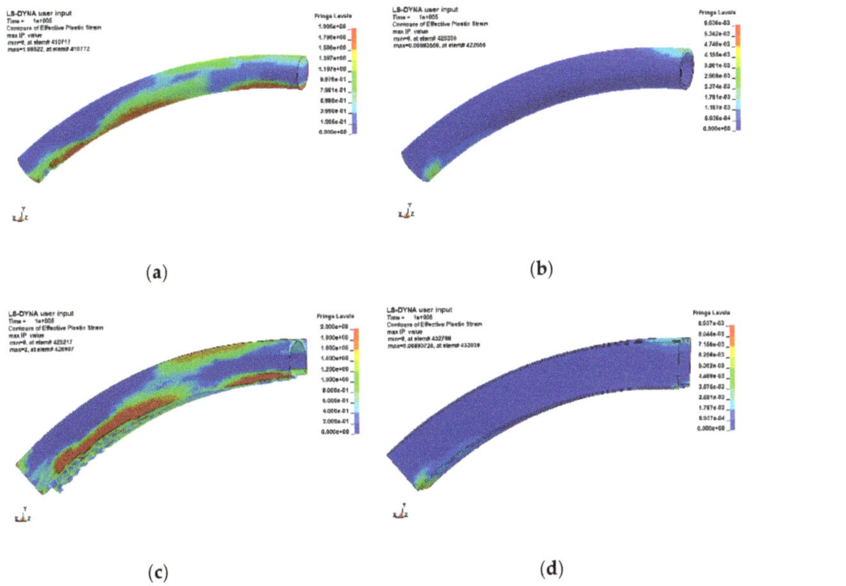

Figure 8. Effective plastic strain distribution of each component of the protective door when ΔP = 10 MPa. (**a**) Inner concrete. (**b**) Steel tube. (**c**) Confined concrete. (**d**) Surface panel and its separators.

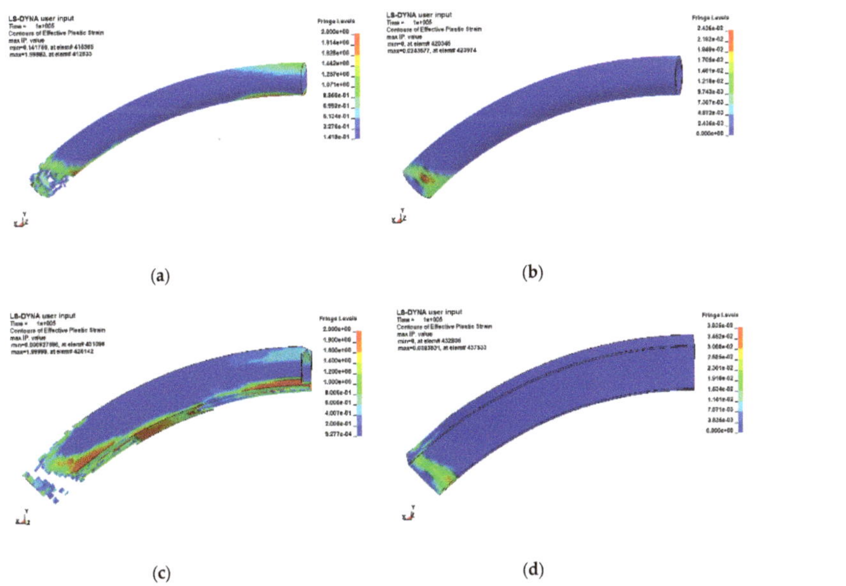

Figure 9. Effective plastic strain distribution of each component of the protective door when ΔP = 12 MPa. (**a**) Inner concrete. (**b**) Steel tube. (**c**) Confined concrete. (**d**) Surface panel and its separators.

The loading time of the two figures above is uniformly taken as the end of the simulation, i.e., 100 ms. Under the blast load of ΔP= 10 MPa, the inner concrete basically does not appear to be damaged. The confined concrete is dislodged on the tensile side, and the steel tube, surface panel, and separators enter the plastic state at the top and the foot of the arch in a part of the section. When ΔP= 12 MPa, the inner concrete and confined concrete at the foot of the arch are fractured, and the steel tube, surface panel, and separators show concentrated plastic strain distributions at the foot of the arch, indicating the trend of the plastic state of the entire section. Under the two different blast loads, the damage mode of the protective door is changed. When ΔP= 10 MPa, the protective door is damaged to some extent at the top and foot of the arch, but the damage limit is not yet reached. When ΔP= 12 MPa, the protective door suffers serious damage, mainly in the arch foot, which is close to the damage limit.

Additional working conditions have been selected, and the description of damage characteristics under each blast load is summarized in Table 1:

Table 1. Damage situation of protective doors under different ΔP.

ΔP	Distribution Diagram of Plastic Strain	Description of Damage Characteristics
7	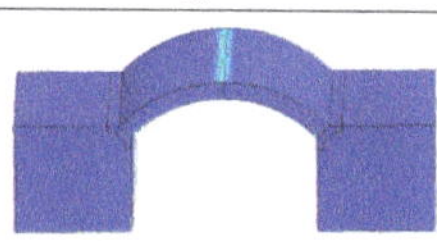	The protective door shows overall deformation response characteristics; the plastic strains appear at the top and foot of the arch, but the distribution of the plastic zone is not large, and the plastic zone at the foot of the arch is smaller.
10	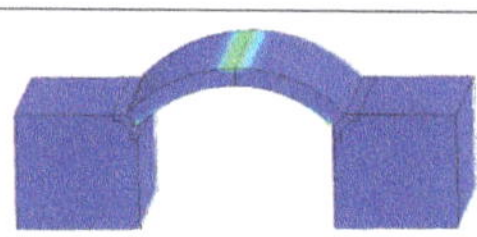	
11	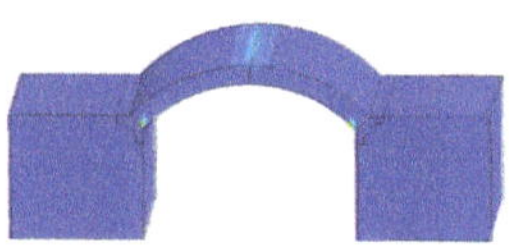	The protective door still shows the overall deformation response characteristics, and the plastic strains appear at both the top and foot of the arch. However, compared to the case of ΔP= 10 MPa, the range of plastic strain distribution at the top of the arch is reduced, while the range of plastic strain at the foot of the arch is increased.
12		The protective door shows local damage characteristics, plastic strains are concentrated at the foot of the arch, the concrete in this position shows "fracture," and the steel is about to enter or has entered the situation of full section yielding.
13	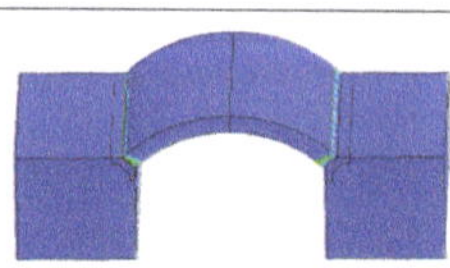	
14	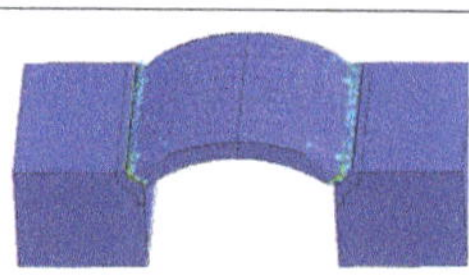	A slip fracture appears at the foot of the arch, and the protective door becomes unstable and collapses.

Table 1 shows that with the increase in ΔP, the degree of damage to the SCCFST arched protective door gradually aggravates. The value of ΔP is less than 11 MPa when the protective door shows the overall deformation response mode in which the load causes a continual increase in damage. This is evidenced by the deformation of the arch foot leading to the slip fracture damage mode. Combined with the description of the dynamic response of the protective door displacement in Section 3.1, it can be deduced that the door becomes dynamically unstable when the "plastic hinge" is formed at the foot of the arch, and the structure becomes a geometrically unstable system that reaches the limit of its load-bearing capacity.

3.3. Damage Mechanism of the Protective Door

According to the analysis results presented in Section 3.2, under large blast loads ($\Delta P \geq 11$ MPa), the regions of plastic damage are mainly located at the foot of the arch. In practical engineering design, it is necessary to consider the situation where the door is subjected to large blast loads, making the arch foot the most susceptible to damage, and the damage limit of the arch determines the load-bearing capacity of the door. According to the yield line theory [23], the plastic hinge formed at the foot of the arch when subjected to a radial blast load uniformly distributed on the outer surface is, by definition, the yield line. The protective door slides and collapses along the yield line and eventually becomes a geometrically unstable system. The diagram of the yield line is as follows:

The damage can be modeled using the maneuver method of limit analysis. If the SCCFST structure is considered an ideal elastic-plastic material, the structure will enter the plastic flow state when the load reaches a certain value. Using the upper limit theorem, the upper limit of the ultimate load can be determined by equating the internal work in the maneuvering tolerance field to the external work. The specific calculation method is shown in Section 4.

4. Theoretical Calculation of Ultimate Blast Resistance-Bearing Capacity of the Protective Door

4.1. Mechanical Model

In the motorized hinge method of limit analysis, each yield line in Figure 10 is integrated to establish the virtual work equation, where the external work performed by the uniform load is given by

$$W = \sum \iint_{A_n} P_i \omega(x, y) \mathrm{d}A_n \tag{1}$$

where, P_i is the load per unit area, $\omega(x, y)$ is the displacement per unit area along the direction of the load, and A_n is the area of the load-acting surface.

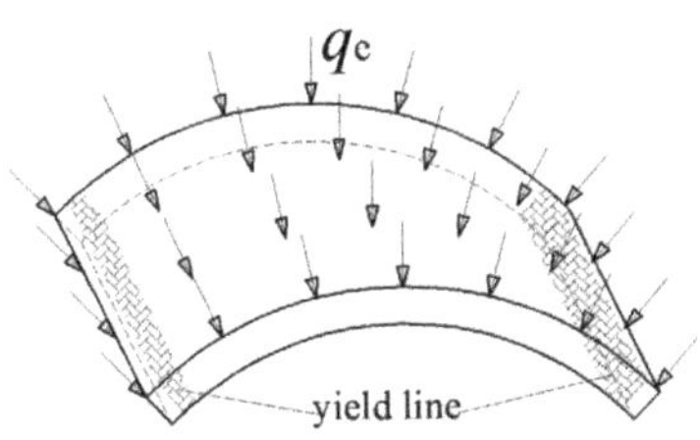

Figure 10. Schematic diagram of yield line.

The internal work performed by the cross-sectional internal force along the yield line is:

$$D = \sum \left[\int_l \alpha_i M_{ui} \mathrm{d}s \right] \tag{2}$$

where, α_i is the normal turning angle, M_{ui} is the unit width of the section limit bending moment, and l is the yield line length, i.e., the section width.

From the geometry,

$$\mathrm{W} = \mathrm{D} \tag{3}$$

According to the principle of imaginary displacement, a normal imaginary displacement δ is applied at each yield line, as shown in Figure 11:

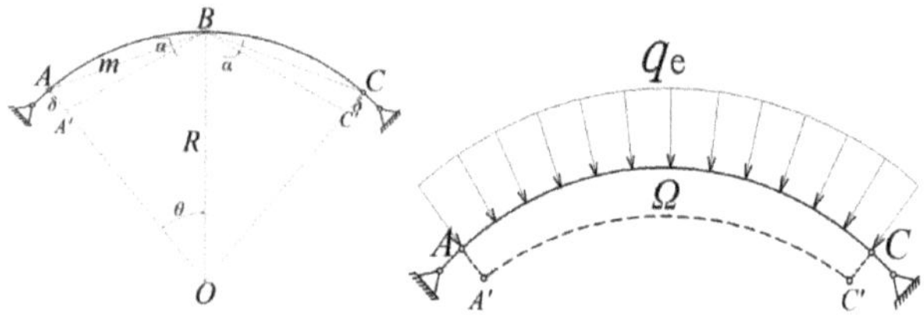

Figure 11. Calculation sketch of the arched protective door.

The bending moment at the two plastic hinges is such that $M_A = M_C = M_u$.
In Figure 11, the initial length of AB is m. From the geometry, it follows that:

$$m = R\sqrt{2(1 - \cos\vartheta)} \tag{4}$$

Since α is very small, it can be approximated as

$$\delta = m \cdot \alpha \tag{5}$$

Then, the work performed by the internal force along the yield line is:

$$D = \sum \alpha_i M_{ui} l_j = 2\alpha M_u l = \frac{2\delta M_u l}{R\sqrt{2(1 - \cos\vartheta)}} \tag{6}$$

The work performed by the external load can be expressed as:

$$W = \sum V_i P_i = q_e \Omega l = q_e l_{AC} \delta \cdot l = 2R\delta\vartheta l \cdot q_e \tag{7}$$

where, Ω is the area formed by the movement of the protective door in the unit width of the cross-section 2-2.

Substituting Equations (6) and (7) into (3), we obtain the expression for the ultimate static load that the protective door can withstand:

$$q_e = \frac{M_u}{R^2\vartheta\sqrt{2(1 - \cos\vartheta)}} \tag{8}$$

To obtain the ultimate blast load that the protective door can withstand, a dynamic coefficient K_d is introduced, which is the ratio of the dynamic effect of the blast load on the protective door to the equivalent static load required to produce the same response in the system. The equivalent static load on the system is:

$$q_e = K_d \cdot \Delta P_m \tag{9}$$

For the elastic-plastic system subjected to the sudden addition of linear decay load shown in Figure 4, the value of K_d is related to the load action time t_1, structural self-oscillation circular frequency ω, and allowable ductility ratio $[\beta]$ as follows [24]:

$$K_d = \left[\frac{2}{\omega t_1}\sqrt{2[\beta] - 1} + \frac{2[\beta] - 1}{4[\beta](1 + 4/\omega t_1)}\right]^{-1} \tag{10}$$

For steel-concrete composite structures, $[\beta]$ is often taken as 3.

In Equation (10), K_d tends to be 1.2 when $\omega t_1 > 8$. The arched protective door considered in this study is consistent with such a situation.

4.2. Ultimate Bending Moment of Cross-Sectional

The ultimate bending moment of the cross-section of the protective door is the positive cross-sectional bending-bearing capacity. It has three components associated with the concrete-filled steel tube, the steel frame that connects the surrounding steel panel and separators, and the confined concrete between the steel tube and steel frame, respectively. Here, the concrete-filled steel tube and separators can be seen as a unified steel bone. The following basic assumptions are satisfied for the calculation [25]:

(1) Steel and concrete have flat sections with respect to deformation;
(2) The plot of stresses in the compressed zone within the concrete can be simplified to an equivalent rectangle;
(3) The plots of tensile and compressive stresses in the steel bones are trapezoidal, which can be simplified to an equivalent rectangle in calculation;
(4) The tensile strength of the concrete is ignored.

The SPSPC section can be divided into (n + 2) subsections or cells as indicated in Figure 2, and the calculation sketch for each sub-section is shown in Figure 12.

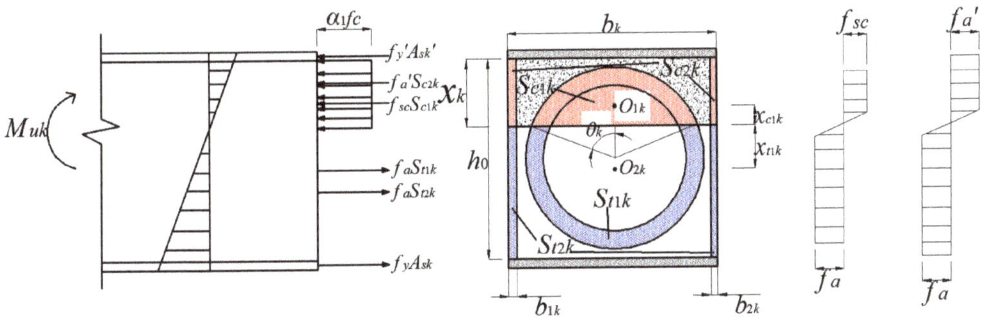

Figure 12. Calculation Sketch of ultimate bending moment of SPSPC section.

According to the force balance:

$$\begin{gathered} \sum N = 0, \\ \alpha_1 f_c b_{eqk} x_k + f_y' A_{sk}' + f_{sc} S_{c1k} + f_a' S_{c2k} = f_a (S_{t1k} + S_{t2k}) + f_y A_{sk} \\ \sum M = 0, \\ M_{uk} = \alpha_1 f_c b_{eqk} \frac{x_k^2}{2} + f_y' A_{sk}' \left(x_k + \frac{h_s}{2} \right) + f_{sc} S_{c1k} x_{c1k} + f_a' S_{c2k} \frac{x_k}{2} \\ + f_a \left(S_{t1k} x_{t1k} + S_{t2k} \frac{h_0 - x_k}{2} \right) + f_y A_{sk} \left(h_0 - x_k + \frac{h_s}{2} \right) \end{gathered} \tag{11}$$

where, α_1 is the concrete strength subfactor. When the strength does not exceed 50 MPa, α_1 is taken as 1.0; when the strength does not exceed 80 MPa, α_1 is taken as 1.0. Elsewhere, α_1 is determined by linear interpolation between 50 MPa and 80 MPa; f_c is the axial compressive strength of concrete; b_{eqk} is the equivalent width of the concrete compression zone of the k calculated section given by

$$b_{eqk} = \frac{b_k x_k - S_{c1k} - S_{c2k}}{x_k} \tag{12}$$

where b_k is the width of the concrete section of the k calculated section; x_k is the height of the concrete compression zone of the k calculated section; $f_y\prime$, f_y are the design values of compressive and tensile strength of the surrounding panel; $A_{sk}\prime$, A_{sk} are the area of the panel in compression and in the tension of the k calculated cross-section; $f_a\prime$, f_a are the design values of the compressive and the tensile strength of the steel bone; f_{sc} is the compressive strength of the combined section of the concrete-filled steel tube.

According to the standard DL/T5085-2021 [26],

$$f_{sc} = \left(1.212 + M\xi + N\xi^2\right) f_c \tag{13}$$

In Equation (11), $M = 0.1759 f_y/235 + 0.974$; $N = -0.1038 f_c/20 + 0.0309$; ξ is the concrete-filled steel tube confinement factor, and $\xi = \frac{f_y S_s}{f_c S_c}$; S_s is the steel tube cross-sectional area in the concrete-filled steel tube; S_c is the concrete cross-sectional area in the concrete-filled steel tube; S_{c1k}, S_{t1k} are the area of the concrete-filled steel tube in the compression zone and tension zone of the kth calculated section.

From the geometry,

$$\begin{aligned} S_{c1k} &= \tfrac{1}{2} r_2^2 (2\theta_k - sin2\theta_k) \\ S_{t1k} &= \left(r_2^2 - r_1^2\right)(\pi - \theta_k) \\ \theta_k &= \cos^{-1} \tfrac{h_0/2 - x_k}{r_2} \end{aligned} \tag{14}$$

S_{c2k}, S_{t2k} are the area of the separators (or surrounding side panel) in the compression zone and tension zone of the kth calculated section;

$$\begin{aligned} S_{c2k} &= (b_{1k} + b_{2k}) x_k \\ S_{t2k} &= (b_{1k} + b_{2k})(h_0 - x_k) \end{aligned} \tag{15}$$

b_{1k}, b_{2k} are the width of steel bone on two sides of the kth calculated section. When $k = 1$, $b_{1k} = h_s$, $b_{2k} = \frac{b_s}{2}$; when $k = n + 2$, $b_{1k} = \frac{b_s}{2}$, $b_{2k} = h_s$; when k is taken as else, $b_{1k} = b_{2k} = \frac{b_s}{2}$; x_{c1k}, x_{t1k} are the distance from the mass center of the concrete-filled steel tube in the compression zone O_{1k} and the mass center of the concrete-filled steel tube in the tension zone O_{2k} to the neutral axis of the kth calculated section.

The ultimate bending moment of the SCCFST section can be derived using the summation formula:

$$M_u = \sum_{k=1}^{n+2} M_{uk} \tag{16}$$

4.3. Analysis of Calculation Results

The ultimate blast resistance-bearing capacity of the protective door can be derived using Equations (8), (9), (11) and (16).

The numerical examples presented in Section 2 are solved using the theoretical methods described in this section.

The f_c of the C60 concrete is 27.5 MPa and α_1 is 0.98, while the $f_y f_y'/f_a/f_a\prime$ of steel is 400 MPa.

In this case, the ultimate blast resistance of the SCCFST arched protective door is 11.28 MPa by the theoretical method. According to the numerical simulation, the ultimate load capacity range obtained is 12–13 MPa, and the results of the theoretical calculation are slightly lower than those of the numerical simulation, but the relative error does not exceed 13%. The reason for this phenomenon is that the theoretical calculation considers only the confinement effect of the steel tube for the concrete inside, whereas the confinement effect of the surrounding panel and its separators for the confined concrete is ignored.

5. Analysis of the Impact of Design Parameters on the Ultimate Blast Resistance-Bearing Capacity

The cross-section of the SCCFST arched protective door consists of a concrete-filled steel tube, a steel frame connected with the surrounding panel and its separators, and a confined concrete section between the steel tube and the steel frame. Changing the design parameters related to any of these components will affect the blast resistance. In this section, the design parameters related to each component are varied. Using the calculation method proposed in Section 4, the effect of these changes on each component of the blast resistance-bearing capacity of the protective door is analyzed.

5.1. *Influence of Concrete-Filled Steel Tube Design Parameters*

From (13), it can be seen that the factors that influence the strength of the combined section of a concrete-filled steel tube include the outer diameter of the steel tube, the wall thickness of the steel tube, the strength of the steel tube, and the strength of the concrete.

For ease of comparison, the dimensional parameters of each component are expressed in terms of the steel ratio (section occupancy) ρ, and it is the percentage of the cross-sectional area of each component to the cross-sectional area of the protective door, which is as follows:

Steel ratio of steel tube:

$$\rho_t = \frac{(n+2)\pi(r_2 - r_1)^2}{[b_1(n+2) + 2(h_s - b_s)](h_1 + 2h_s)} \tag{17}$$

Occupancy of concrete-filled steel tube section:

$$\rho_{sc} = \frac{(n+2)\pi {r_2}^2}{[b_1(n+2) + 2(h_s - b_s)](h_1 + 2h_s)} \tag{18}$$

Steel ratio of surrounding panel and separators:

$$\rho_f = \frac{2h_s[b_1(n+2) + 2(h_s - b_s) + h_1] + (n+1)b_s h_1}{[b_1(n+2) + 2(h_s - b_s)](h_1 + 2h_s)} \tag{19}$$

Since the outer diameter and wall thickness of the steel tube determine the geometric dimension of internal concrete, the section occupancy of the pipe ρ_{sc} and the steel ratio of the pipe ρ_t are analyzed first.

At ρ_t= 19.42%, the corresponding relationship between different section occupancies of the concrete-filled steel tube and the blast resistance of the protective door is shown in Figure 13. At ρ_{sc} = 61.14%, the corresponding relationship between the different section occupancies of the steel tube and the blast resistance of the protective door is shown in Figure 14.

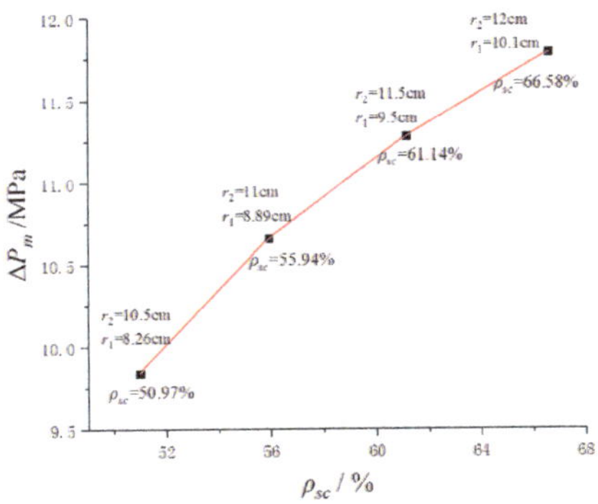

Figure 13. Effect of ρ_{sc} on the protective door load-bearing capacity.

The relationship between the strength of the steel tube and the load-bearing capacity of the protective door is shown in Figure 15, and the relationship between the strength of the concrete inside the steel tube and the load-bearing capacity of the protective door is shown in Figure 16.

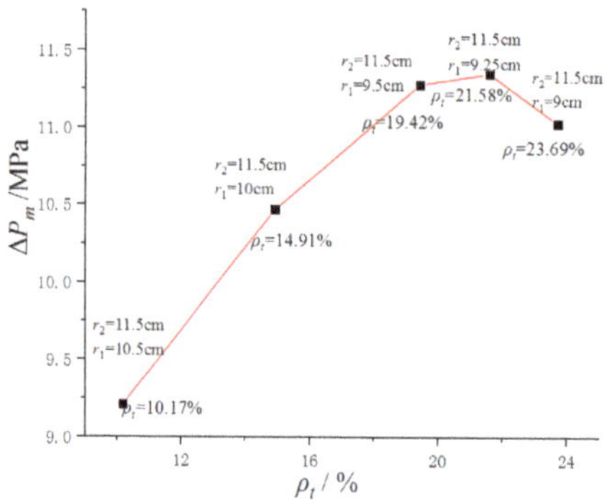

Figure 14. Effect of ρ_t on the protective door load-bearing capacity.

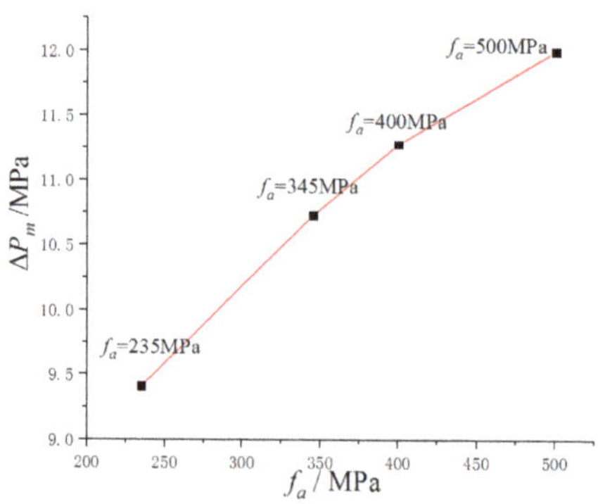

Figure 15. Effect of steel tube strength on the protective door load-bearing capacity.

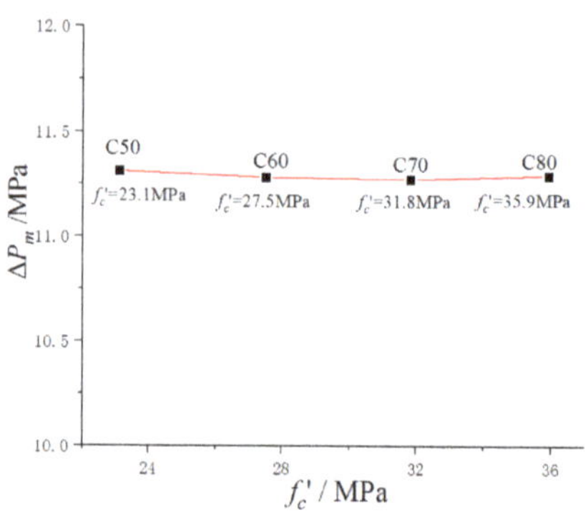

Figure 16. Effect of the strength of the concrete in the steel tube on the protective door load-bearing capacity.

From Figures 13–16, it can be observed that an increase in the section occupancy of the concrete-filled steel tube and steel tube strength can improve the blast load-bearing capacity of the protective door, whereas an increase in the steel ratio of the concrete-filled steel tube strength inside the steel tube in a certain interval has a negative effect on the blast load-bearing capacity of the protective door. This is because the confinement coefficient is an important factor that affects the strength of the combined structure when the concrete-filled steel tube is considered a composite structure. According to Equation (13), if the confinement coefficient is too large, the composite structure of the concrete-filled steel tube cannot perform as well, and it will result in a decrease in strength. Therefore, the wall thickness and strength of the steel tube and the concrete strength should be controlled within the appropriate range. Without changing the steel ratio, in comparison, the outer diameter of the steel tube can be increased to improve the blast resistance. This approach can make full use of the confinement of the steel tube on the concrete and offers an economical solution. Hence, it is advisable to maximize the increase in the outer diameter of the steel tube.

5.2. Influence of Design Parameters of Confined Concrete

The geometry of the confined concrete changes with that of the steel tube, surrounding panel, and separators. Thus, the effect of the geometry of confined concrete on the blast resistance is not considered in this subsection, and only the effect of changes in the concrete strength is analyzed, as shown in Figure 17.

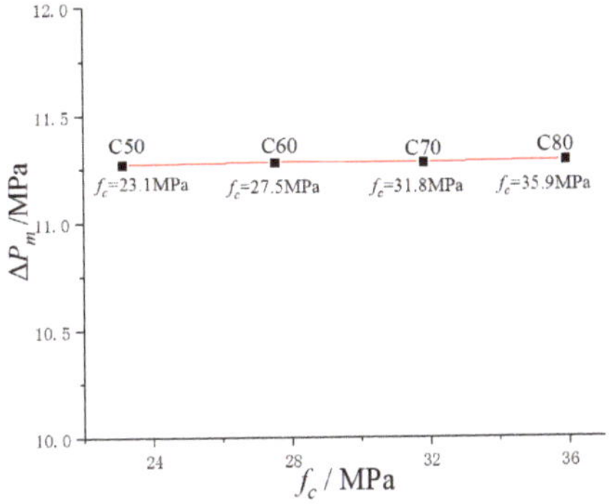

Figure 17. Effect of the strength of confined concrete on the protective door load-bearing capacity.

It can be seen from Figure 17 that as the strength grade of the confined concrete increases from C50 to C80, the load-bearing capacity of the protective door also rises slowly, but ΔP_m has been maintained at about 11.3 MPa, and the change is not obvious.

5.3. Influence of Design Parameters of the Surrounding Panel and Its Separators

The variable design parameters of the surrounding panel and separators are the steel ratio and steel strength, which do not change the thickness of the separators. The steel ratio is changed by adjusting the thickness of the surrounding panel. The influence of these two parameters on the blast resistance of the protective door is illustrated in Figures 18 and 19.

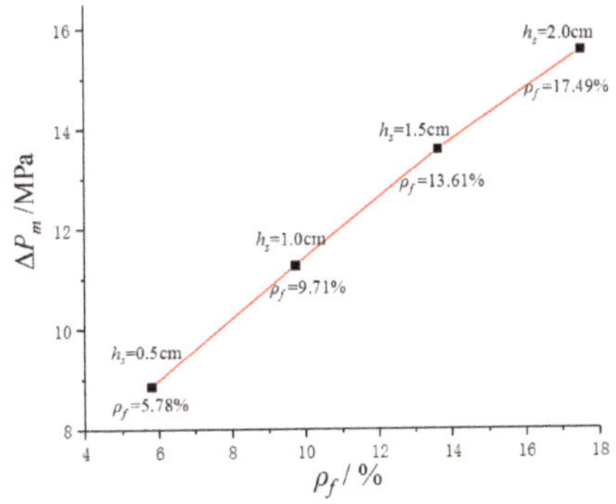

Figure 18. Effect of ρ_f on the protective door load-bearing capacity.

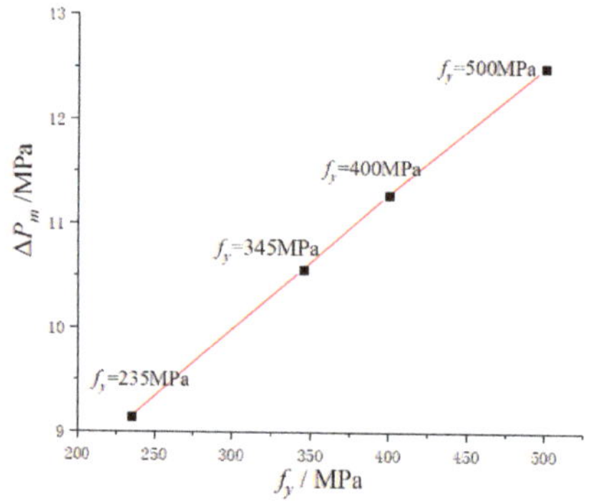

Figure 19. Effect of steel strength on the protective door load-bearing capacity.

From the two figures above, it can be seen that ΔP_m increases by 75.5% when the steel ratio ρ_f increases by 11.71% and by 36.8% when the steel strength is nearly doubled.

6. Conclusions

The following conclusions can be drawn from the study described herein:

(1) Under a radially uniform load on the outer surface, as the load increases, the SCCFST arched protective door supported on both sides shifts from the overall deformation reaction mode into the development of a plastic hinge line at the foot of the two arches, resulting in dynamic instability. The protective door becomes a geometrically unstable system and eventually fails due to slip fracture damage.

(2) This paper proposes a method for calculating the ultimate blast resistance of the SCCFST arched protective door based on the motorized hinge method of ultimate analysis. This method considers the geometric relationship between the "plastic hinge" damage mode of the door and the coordinated action of the components of the door section and can be used for the design and analysis of protective doors against blast load.

(3) The impact of the design parameters of each part of the SCCFST arched protective door on the ultimate blast resistance-bearing capacity has a clear pattern: (A) The primary factor influencing the ultimate blast resistance is the surrounding panel and its separators. Hence, increasing the steel ratio and steel strength can significantly increase the ultimate blast resistance load capacity. (B) The concrete-filled steel tube contributes to the ultimate blast resistance to a lesser extent. Enhancing the steel ratio and strength of the steel tube within a certain range can moderately improve the ultimate blast resistance. (C) While confined concrete plays a role in the ultimate blast resistance capacity, the impact of increasing the concrete class on the load-bearing capacity is not particularly pronounced.

Author Contributions: Conceptualization, S.D. and Z.T.; methodology, S.D. and Z.T.; software, S.D.; validation, X.C., C.T. and Z.W.; formal analysis, C.T.; investigation, C.T.; resources, X.C.; data curation, Z.W.; writing—original draft preparation, S.D.; writing—review and editing, Z.T.; visualization, X.C.; supervision, Z.W.; project administration, Z.T.; funding acquisition, Z.T. All authors have read and agreed to the published version of the manuscript.

Funding: This research received no external funding.

Data Availability Statement: The raw/processed data required to reproduce these findings cannot be shared at this time as the data also form part of an ongoing study.

Conflicts of Interest: The authors declare no conflict of interest.

References

1. Jacques, E.; Lloyd, A.; Berry, T.; Saatcioglu, M.; Shinder, J. Development of blast resistant steel doors. In Proceedings of the 11th International Conference on Shock & Impact Loads on Structures, Ottawa, ON, Canada, 14–15 May 2015; pp. 14–15.
2. Meng, Y.; Li, B.; Wang, Y. Structure design of new airtight blast door based on topology and shape optimization method. *Geotech. Geol. Eng.* **2016**, *34*, 703–711. [CrossRef]
3. Veeredhi, L.S.B.; Ramana Rao, N.V. Studies on the impact of explosion on blast resistant stiffened door structures. *J. Inst. Eng. Ser. A* **2015**, *96*, 11–20. [CrossRef]
4. Thimmesh, T.; Shirbhate, P.; Mandal, J.; Sandhu, I.; Goel, M. Numerical investigation on the blast resistance of a door panel. *Mater. Today Proc.* **2021**, *44*, 659–666. [CrossRef]
5. He, H.; Zhang, B.; Zheng, Q.; Fan, H. Anisotropic dynamic theory to predict blast responses of composite fluted-core sandwich protective door panels. *Thin-Walled Struct.* **2021**, *161*, 107436. [CrossRef]
6. Li, X.; Miao, C.; Wang, Q.; Geng, Z. Antiknock performance of interlayered high-damping-rubber blast door under thermobaric shock wave. *Shock. Vib.* **2016**, *2016*, 2420893. [CrossRef]
7. Zhang, B.; Chen, H.; Zhao, Z.; Fan, H.; Jin, F. Blast response of hierarchical anisogrid stiffened composite panel: Considering the damping effect. *Int. J. Mech. Sci.* **2018**, *140*, 250–259. [CrossRef]
8. Meng, F.; Zhang, B.; Zhao, Z.; Xu, Y.; Fan, H.; Jin, F. A novel all-composite blast-resistant door structure with hierarchical stiffeners. *Compos. Struct.* **2016**, *148*, 113–126. [CrossRef]
9. Zhang, Y.; Fan, J.; Tao, J.; Jia, X.; Huang, Z.; Liu, X.; Liang, S.; Zhao, Y.; Bai, X. A sutdy on impact resistance performance of a blast door with POZD coating. *J. Vib. Shock.* **2022**, *4*, 231–238. (In Chinese)
10. Shi, S.; Zhang, X.; Yin, P. Static Analysis of the New Defensive Structure under Explosive Loading. *Undergr. Space* **2003**, *23*, 66–68+109. (In Chinese)
11. Ganorkar, K.; Goel, M.D.; Chakraborty, T. Numerical Analysis of Double-Leaf Composite Stiffened Door Subjected to Blast Loading. *J. Perform. Constr. Facil.* **2023**, *37*, 04022067. [CrossRef]
12. Tan, Z. *The Dynamic Response and Optimum Design of Vaulted Blast Resistant Door*; Chang'an University: Xi'an, China, 2015.
13. Manojkumar, V.C.; Mattur, C.N.; Kulkarni, S.M. Axial strength of circular concrete-filled steel tube columns-DOE approach. *J. Constr. Steel Res.* **2010**, *66*, 1248–1260.
14. Chung, K.; Park, S.; Choi, S. Material effect for predicting the fire resistance of concrete-filled square steel tube column under constant axial load. *J. Constr. Steel Res.* **2008**, *64*, 1505–1515. [CrossRef]
15. Jiang, Z.; Sun, S.; Lu, H.; Yu, S. Analysis of dynamic response and influence factors of concrete-filled steel and steel tube blast doors. *J. Ordnance Equip. Eng.* **2022**, *43*, 41–47. (In Chinese)
16. Li, E.; Sheng, X.; Wang, J. Dynamic analysis of vaulted protective doors under nuclear and conventional blast loadings. *Prot. Eng.* **2011**, *33*, 16–21. (In Chinese)
17. Chen, W.; Hao, H. Numerical study of a new multi-arch double-layered blast-resistance door panel. *Int. J. Impact Eng.* **2012**, *43*, 16–28.
18. Li, C.; Qin, F.; Zhang, Y.-D.; Zhang, Y.; Fan, J.-Y. Numerical and Experimental Investigations on the blast-resistant properties of arched RC blast doors. *Int. J. Prot. Struct.* **2010**, *1*, 425–441.
19. Guo, D.; Li, Z.; Wang, Q.; Hou, X. Research on blast-resistant properties of concrete-filled steel and steel tube blast doors. *Prot. Eng.* **2013**, *35*, 38–43. (In Chinese)
20. *LS-DYNA Keyword User's Manual*; LSTC: San Francisco, CA, USA, 2017.
21. Xu, Y.; Hu, S. A study on the dynamic ultimate capacity of CFST arch bridges. *China Civ. Eng. J.* **2006**, *39*, 68–73. (In Chinese)
22. Budiansky, B.; Roth, R.S. Axisymmetric Dynamic Buckling of Clamped Shallow Spherical Shells. *NASA TN* **1962**, *510*, 597–606.
23. Shen, J.; Wang, C.; Jiang, J. *Finite Element of Reinforced Concrete and Limit Analysis of Shell*; Tsinghua University Press: Beijing, China, 1993.
24. US Department of the Army. *TM5-1300 Structures to Resist the Effects of Accidental Explosions*; US Department of the Army: Washington, DC, USA, 1990.
25. Ma, H.; Wang, T. *Combination Structure of Steel-Concrete*; Building Materials Industry Press: Beijing, China, 2006.
26. *DL/T5085-2021*; Code for Design of Steel-Comcrete Composite Structure. Department of Transportation: Beijing, China, 2021.

MDPI
St. Alban-Anlage 66
4052 Basel
Switzerland
www.mdpi.com

Buildings Editorial Office
E-mail: buildings@mdpi.com
www.mdpi.com/journal/buildings